KU-567-027

TELECOMMUNICATION NETWORKS AND COMPUTER SYSTEMS

Series Editors

Mario Gerla
Aurel Lazar
Paul Kühn
Hideaki Takagi

#27431047

Haruo Akimaru Konosuke Kawashima

Teletraffic

Theory and Applications

With 81 Illustrations

Springer-Verlag
London Berlin Heidelberg New York
Paris Tokyo Hong Kong
Barcelona Budapest

Haruo Akimaru, Professor
Asahi University, Hozumi-cho, Motosu, Gifu Prefecture, Japan 501-02
Toyohashi University of Technology, Tempaku-cho, Toyohashi, Japan 441
Information Networks Research Institute, *INR*, Tokyo, Montreal, Toyohashi

Konosuke Kawashima
NTT Telecommunication Networks Laboratories,
3-9-11 Midori-cho, Musashino-shi, Japan 180

Series Editors

Mario Gerla
Department of Computer Science
University of California
Los Angeles
CA 90024, USA

Paul Kühn
Institute of Communications
Switching and Data Technics
University of Stuttgart
D-7000 Stuttgart, Germany

Aurel Lazar
Department of Electrical Engineering and
Center for Telecommunications Research
Columbia University
New York, NY 10027, USA

Hideaki Takagi
IBM Japan Ltd
Tokyo Research Laboratory
5-19 Sanban-cho
Chiyoda-ku, Tokyo 102, Japan

ISBN 3-540-19805-9 Springer-Verlag Berlin Heidelberg New York
ISBN 0-387-19805-9 Springer-Verlag New York Berlin Heidelberg

British Library Cataloguing in Publication Data
A catalogue record for this book is available from the British Library

Library of Congress Cataloging-in-Publication Data
A catalog record for this book is available from the Library of Congress

Apart from any fair dealing for the purposes of research or private study, or criticism or review, as permitted under the Copyright, Designs and Patents Act 1988, this publication may only be reproduced, stored or transmitted, in any form or by any means, with the prior permission in writing of the publishers, or in the case of reprographic reproduction in accordance with the terms of licences issued by the Copyright Licensing Agency. Enquiries concerning reproduction outside those terms should be sent to the publishers.

© H. Akimaru and K. Kawashima 1993
Printed in Germany

The publisher makes no representation, express or implied, with regard to the accuracy of the information contained in this book and cannot accept any legal responsibility or liability for any errors or omissions that may be made.

Typesetting: Camera ready by authors
69/3830-543210 Printed on acid-free paper

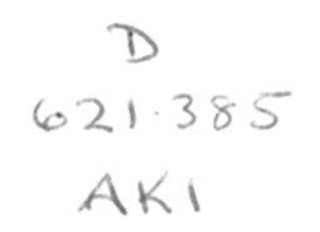

PREFACE

Telecommunications systems have been evolving from the conventional telephone network that mainly deals with voice, to the ISDN (integrated services digital network) integrating voice, data and image. Moreover, the ATM (asynchronous transfer mode) and optical switching technologies are being developed for the broadband ISDN which can handle the high speed video communications as well. Computer networks are also progressing from centralized TSS (time-sharing system) to distributed LAN (local area network) and VAN (value added network).

In the research, development, design and operation of such telecommunications and computer networks, the important problems are determining the optimum configuration and dimensions of the systems for providing a given performance or GOS (grade of service). The teletraffic theory, the basis for the performance evaluation and the dimensioning, has been studied along with the switching technology, and has developed rapidly by incorporating the recent advances in OR (operations research) and queueing theory. However, it is sometimes difficult for non-experts of teletraffic to understand and apply these theories, because they require a deep mathematical background.

This book is translated and extended from the Japanese version originally published in 1990 by Telecommunications Association of Japan. Its aim is to allow researchers, engineers and managers of the teletraffic and computer systems to apply the latest theories in practice. The theory for the basic models is described in detail, while for advanced models requiring a more advanced theory, practical formulas are given, and detailed derivations are left to the references. Examples explain how the theories are applied to the actual systems, and exercises are included for self-study.

Chapter 1 gives examples of teletraffic systems, the definitions required for modeling the systems, and fundamental relations useful for theoretical analysis of the models. Chapter 2 describes the traditional analyses for basic (Markovian) models. Chapters 3 and 4 present the main results for more sophisticated (non-Markovian and multi-class input) models. Chapter 5 deals with the alternative routing system which has been widely applied in the telephone network. Chapter 6 gives the analyses and approximations for advanced systems including the ATM and LAN.

There are a number of systems for which analytical solutions have not yet been obtained. Chapter 7 describes the computer simulation that can be used to evaluate the performance of such systems. The simulation can also be used to verify the approximate solutions for complex systems. The bibliography contains only references used in this book, although there are a lot of other literature on teletraffic theory and engineering. The appendices include fundamental formulas, the BASIC program lists, a basis of probability theory, and solutions to the exercises.

This book has been written both as a text for college and university students, and as a handbook for practitioners. In order to understand the basic theories, the reader is recommended to read through Chapters 1,2 and 5. Ones interested in more advanced theories may proceed through Chapters 1,3 and 4. Chapters 6 and 7 are self-contained, respectively, and can be addressed directly by the readers interested in the ATM and LAN, as well as the computer simulation.

The advanced information network will be developed as an infrastructure to meet the social demands, introducing new technologies and systems. It is hoped that this book will contribute to construction of optimum systems by promoting the application of the teletraffic engineering.

The authors wish to express their sincere appreciation to Professor Paul Kuehn at University of Stuttgart for his encouragement and review of this book, and the authors of the references cited. Many thanks are also extended to T. Okuda, Z.S. Niu, T. Kawai, T. Kokubugata, and other graduate students at the Toyohashi University of Technology, and H. Yoshino and Y. Takahashi at the NTT Network Traffic Laboratory, for their help in preparing this English version.

July 1992.

H. Akimaru

K. Kawashima

Contents

Chapter 1

INTRODUCTION

This chapter describes features of teletraffic systems. Definitions and modeling of the systems are presented as well as fundamental relations useful for the analysis of teletraffic systems.

The basis of the probability theory needed for understanding this book is summarized in Appendix C.

1.1 Features of Teletraffic Systems

In telecommunications systems, it would be uneconomical if exclusive resources such as switching and transmission facilities were dedicated to each customer. Therefore a pool of facilities is provided in common for a number of customers, and thus situations can arise in which a customer is rejected or has to wait for connection, because of shortage of the common resources. Hence, it is required to evaluate such grades of services (GOS) quantitatively, and clarify relations between the GOS and the amount as well as configuration of the telecommunications facilities. The *teletraffic theory* analyses such GOS using probability theory since the demand of telephone and data calls has stochastic characteristics. The teletraffic theory has been developing along with telephone switching systems since it was created by A. K. Erlang, a Danish mathematician, at the beginning of the 20th century.

In the field of operations research, the *queueing theory* which deals with similar problems, has also been studied and applied to a wide range of areas such as queueing systems, vehicular traffic, inventory, process control, etc. For recent computer systems, the analysis of *queueing networks* has also been studied. Integrating these related results, the teletraffic theory has been established, and the teletraffic engineering applying the theory plays an important role in practical applications.

Some examples on teletraffic issues are given below. Suppose that we are requested to determine the number s of trunk circuits between two PBXs (private branch exchanges) each accommodating 1000 telephone sets, as shown in Figure 1.1. In order to guarantee that all telephone sets accommodated in the separated PBXs to talk each other simultaneously, the number required would be $s = 1000$. However, this seems too wasteful because such an extreme request would rarely occur. At the other extreme, if $s \doteq 1$, the service would be insufficient. Applying the teletraffic theory, it will be shown that if each telephone set is used at random once an hour for 6 minutes on average, $s = 64$ guarantees the connections with probability 99%. Example 2.1 in Chapter 2 will describe this model.

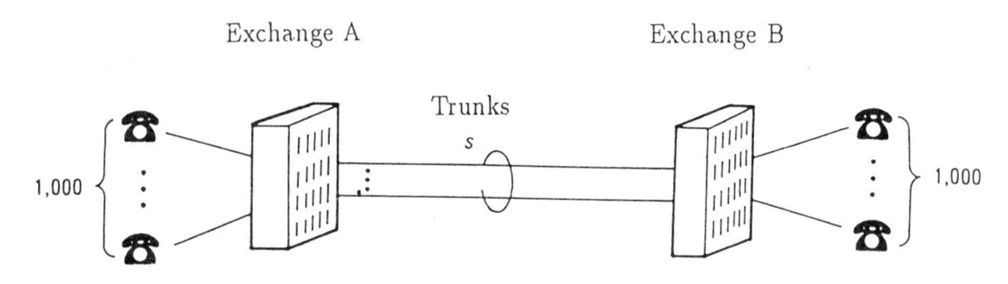

Fig.1.1 Example of teletraffic system

Next, consider a data transmission system as shown in Figure 1.2, where fixed length packets of 2400 bits arrive at the rate of 3 packets per second at random. The question might arise: Which is better in GOS (a) 4 lines of 2400 b/s or (b) a single line of 9600 b/s? The answer will be found in Examples 3.5 and 3.6 in Chapter 3.

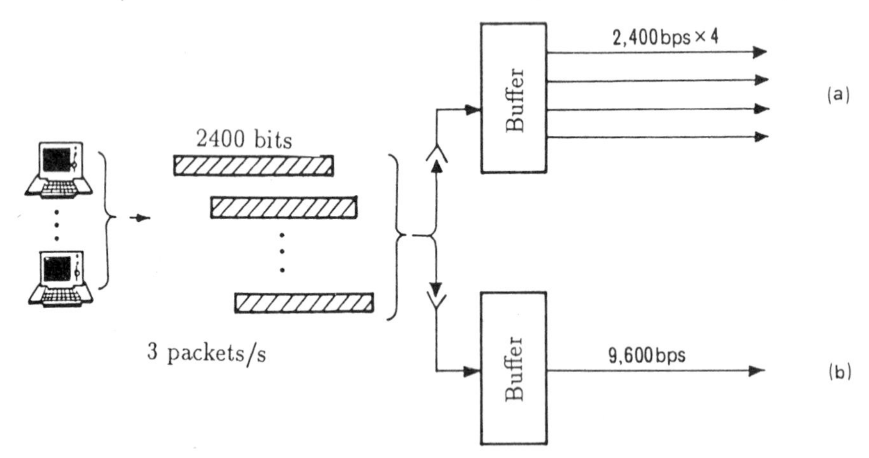

Fig.1.2 Packet transmission system

In order to make such performance evaluations, various factors such as quantity of demand, call origination pattern and type of service, must be specified, which will be described in the next section.

1.2 Modeling of Teletraffic Systems

1.2.1 Traffic Load

A *call* is defined as a demand for a connection in teletraffic systems, and it is also referred to as *customer*. The *holding time* is defined as the duration of the call, which is also called *service time*. The *traffic load* is defined as the total holding time per unit time. The unit of traffic load is named *erlang* (erl) after the creator of the teletraffic theory.

[Example 1.1] Suppose that there are 3 calls per hour with holding times of 5, 10 and 15 minutes, respectively, as shown in Figure 1.3. Then the traffic load a is calculated as

$$a = \frac{(5 + 10 + 15)\,\text{min}}{60\,\text{min}} = 0.5\,\text{erl}.$$

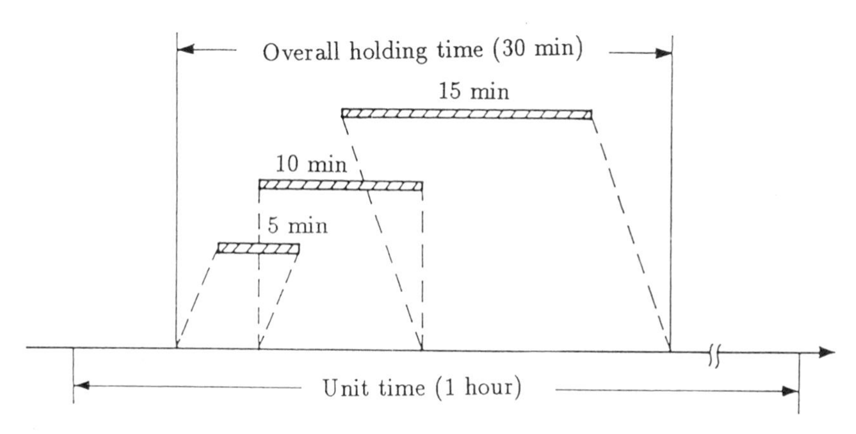

Fig.1.3 Definition of traffic load

The traffic load is sometimes referred to as *traffic intensity*, and has the following properties:

(1) Let c be the number of calls originating per unit of time, and h the mean holding time. Then, the traffic load a is given by

$$a = ch \ \ [\text{erl}]. \tag{1.1}$$

(2) The traffic load is equal to the number of calls originating in the mean holding time.

(3) The traffic load carried by a single trunk is equivalent to the probability (fraction of time) that the trunk is used (busy).

(4) The traffic load carried by a group of trunks is equivalent to the mean (expected) number of the busy trunks in the group.

Properties (1) to (3) are understood readily from the definition of the traffic load. Property (4) may be interpreted as follows: Suppose that a group with s trunks carries the traffic load a erl. Then, the load carried by a trunk is $a_1 = (a/s)$ erl on average, which is equivalent to the probability that the single trunk is busy from Property (3). Hence, the mean (expected) number of busy trunks is given by the relation of the number multiplied by the probability, *i.e.* $sa_1 = a$.

1.2.2 Call Origination Process

As will be seen later, there are various patterns of call origination. Here, let us consider random origination, which is modeled as, as $\Delta t \to 0$,

(1) The probability that a call originates in time interval $(t, t+\Delta t]$ tends to $\lambda \Delta t$ independent of t, where λ is a constant.

(2) The probability that two or more calls originate in $(t, t+\Delta t]$ tends to zero.

(3) Calls originate independently of each other.

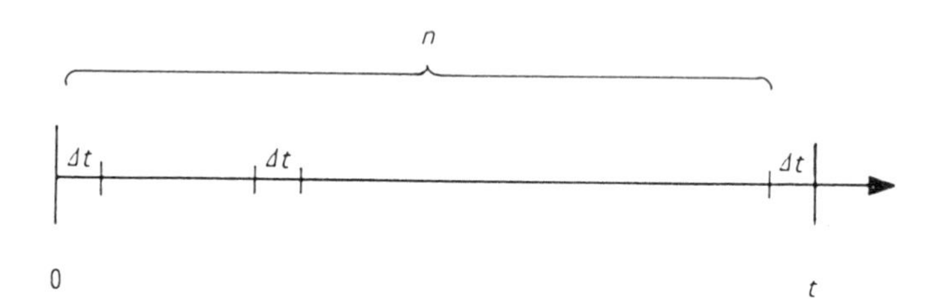

Fig.1.4 Model of call origination

In this model, let us calculate the probability $p_k(t)$ that k calls originate in time $(0, t]$. As shown in Figure 1.4, partition the interval $(0, t]$ into a sufficiently large number n of subsections and let $\Delta t = t/n$. Then, since the probability that exactly k calls originate in k particular subsections is given by $(\lambda \Delta t)^k (1-\lambda \Delta t)^{n-k}$, as $\Delta t \to 0$, and there are $\binom{n}{k}$ exclusive such occurrences, we have

$$
\begin{aligned}
p_k(t) &= \lim_{n\to\infty} \binom{n}{k} \left(\frac{\lambda t}{n}\right)^k \left(1 - \frac{\lambda t}{n}\right)^{n-k} \\
&= \lim_{n\to\infty} \frac{(\lambda t)^k}{k!} \left(1 - \frac{\lambda t}{n}\right)^{n-k} \frac{n}{n}\frac{n-1}{n} \cdots \frac{n-k+1}{n}
\end{aligned}
$$

$$= \frac{(\lambda t)^k}{k!}e^{-\lambda t}. \tag{1.2}$$

This is the *Poisson distribution* with mean λt, where λ is called the *arrival rate* or *origination rate.* (See Table C.1 in Appendix C.) The fact that λ is constant and independent of time is one feature of the random origination, and this model is also referred to as *Poisson arrival* (- *call,* - *input,* - *origination,* - *process*, etc.).

Since the mean number of calls originated in $(0, t]$ is λt, λ is interpreted as the mean number of arrivals in unit time, which is the mean of c in (1.1). The arrival rate depends on the unit of time used, and if the hour is used, it is measured in BHCA (busy hour call attempts).

From (1.2), the probability that no calls originate in $(0, t]$ is given by

$$p_0(t) = e^{-\lambda t}. \tag{1.3}$$

Hence, the distribution function of the interarrival time (probability that the interarrival time is no greater than t) is given by

$$A(t) = 1 - e^{-\lambda t} \tag{1.4}$$

which is the *exponential distribution* with mean λ^{-1}. (See Table C.2 in Appendix C.) Thus, the exponentially distributed interarrival time is another feature of the random origination.

1.2.3 Service Time Distribution

Next, let us consider the distribution of service time. In the simplest case, it is assumed that a call is terminated at random. Taking the origin at the call origination instant, the probability that the call is terminated in $(t, t + \Delta t]$ is $\mu\Delta t$ independent of t, from the random terminating assumption. The complementary distribution function $H(t)$ (probability that the service time is greater than t) is equal to the probability that the call is not terminated in $(0, t]$. Partitioning $(0, t]$ into a sufficiently large number n of subsections and setting $\Delta t = t/n$, then since the latter probability equals $(1 - \mu\Delta t)^n$, as $n \to \infty$, $H(t)$ is given by

$$H(t) = \lim_{n\to\infty} \left(1 - \frac{\mu t}{n}\right)^n = e^{-\mu t}. \tag{1.5}$$

Thus, the service time is exponentially distributed with mean μ^{-1}, where μ is called the *service rate* or *termination rate.* This is often referred to as *exponential service time* in short, and the traffic load is expressed as $a = \lambda/\mu$ from (1.1).

The assumption of exponential service time agrees fairly well with actual telephone conversation times as shown in Figure 1.5. Furthermore, because of the simplicity for theoretical analysis as shown later, it has been widely used in telephone traffic theory.

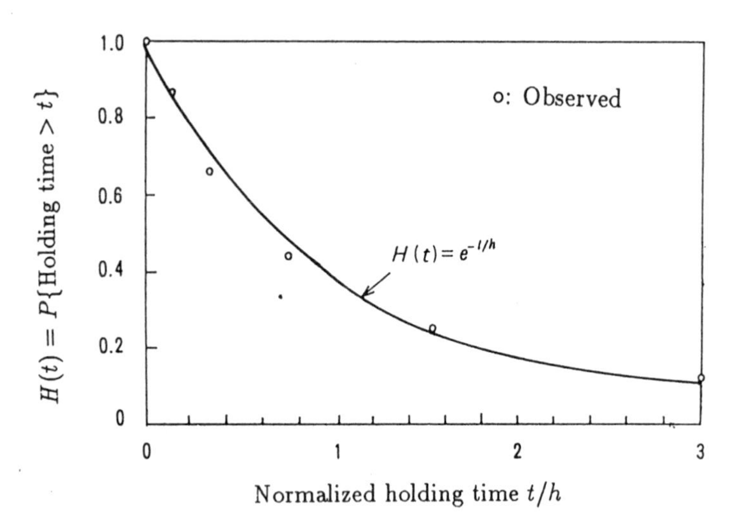

Fig.1.5 Distribution of telephone conversation

1.2.4 Classification of Traffic Models

A system connecting between inlets and outlets is called a *switching system*. If any inlet can be connected to any idle outlet, it is called a *full availability system*; otherwise a *limited availability system*. The condition that the connection can not be made because of busy outlets or internal paths of the switching system is called *congestion*. In congestion, if an incoming call is blocked, the system is called a *loss* or *non-delay system*. If the call can wait for connection, it is called a *waiting* or *delay system*. These systems are illustrated as in Figure 1.6, for example.

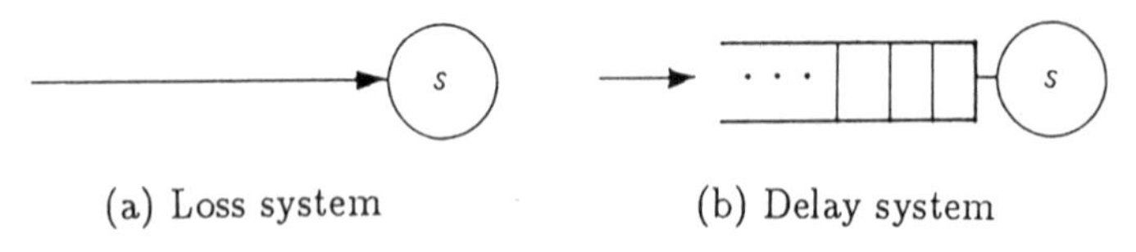

(a) Loss system (b) Delay system

Fig.1.6 Symbols for traffic models

Full availability systems are described by the following:

(1) *Input process* describes ways of call origination or arrival. Although the Poisson process was shown earlier, there are various other processes that will be discussed later.

(2) *Service mechanism* describes the number of outlets (trunks, servers, etc.), holding (service) time distribution, etc. As stated, the exponential distribution applies to telephone traffic, but other distributions (for example constant or deterministic) may be required for data packets.

(3) *Queue discipline* specifies ways for handling calls during congestion (loss or delay). In delay systems, the order of serving waiting calls is to be specified, such as *first-in first-out* (FIFO), *last-in first-out* (LIFO), *random service order* (RSO), etc.

In order to classify full availability systems, *Kendall notation*,

$$\mathrm{A}/\mathrm{B}/s \tag{1.6}$$

is used, where A represents distribution of input process, B service time distribution, and s the number of servers. For A and B, the following symbols are used:

M : Exponential (Markov)

E_k : Phase k Erlangian (convolution of k exponentials with identical mean)

H_n : Order n hyper-exponential (alternative of n exponentials)

D : Deterministic (fixed)

G : General (arbitrary)

GI : General independent (renewal)

MMPP : Markov modulated Poisson process (non-renewal)

For example, a system with Poisson arrival, exponential service time and s servers is expressed as $\mathrm{M}/\mathrm{M}/s$. With finite n inlets (sources), we have $\mathrm{M}(n)/\mathrm{M}/s$, and with waiting room of m positions, $\mathrm{M}/\mathrm{M}/s(m)$ or $\mathrm{M}/\mathrm{M}/s/s+m$. Therefore, a loss system is expressed as $\mathrm{M}/\mathrm{M}/s(0)$ or $\mathrm{M}/\mathrm{M}/s/s$. Without such additional notations, an FIFO delay system with infinite waiting room is meant.

1.3 Fundamental Relations

1.3.1 Markov Property

Consider the duration time X of a phenomenon, say service time, and take the origin at its origination as shown in Figure 1.7. If X is exponentially distributed with mean μ^{-1}, the probability that the phenomenon continues after time instant $x$is given by

$$P\{X > x\} = e^{-\mu x}. \tag{1.7}$$

Hence, the conditional probability that the phenomenon continues further for time period t, given it has lasted until time x, is calculated as

$$\begin{aligned} P\{X > x + t \mid X > x\} &= \frac{P\{X > x + t\}}{P\{X > x\}} = \frac{e^{-\mu(x+t)}}{e^{-\mu x}} \\ &= e^{-\mu t} = P\{X > t\}. \end{aligned} \tag{1.8}$$

It should be noted that the last probability in (1.8) is independent of time x. This implies that the stochastic behavior of the phenomenon after (*future*) time x is only dependent on the state at time x (*present*) and independent of the progress before (*past*) time x. This characteristic is called the *Markov property* or *memoryless property*, and it is known that only the exponential distribution has this property in continuous distributions.

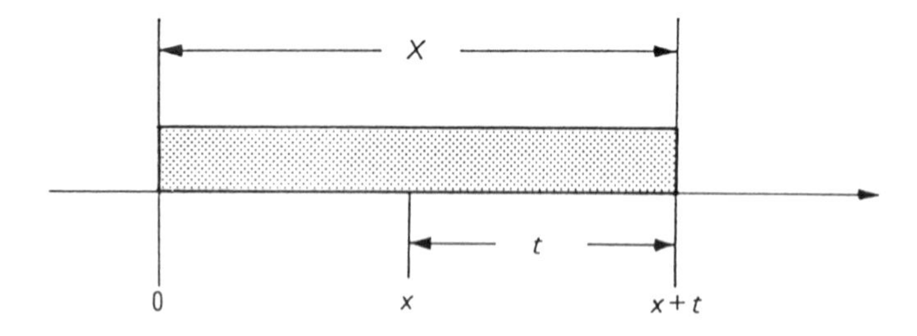

Fig.1.7 Markov property

In fact, if X is exponentially distributed, the *residual time* seen at an arbitrary time instant is also exponentially distributed just the same as the *lifetime* (original duration) X. In what follows, a model with interarrival time and service time both exponentially distributed is called a *Markovian model*; otherwise it is called a *non-Markovian model*.

1.3.2 PASTA

Let P_j be the probability that j calls exist at an arbitrary instant in steady state, and Π_j the corresponding probability just prior to call arrival epoch, as shown in Figure 1.8. Then, in general, these two probabilities are not equal. However, for a system with exponential interarrival time (Poisson arrival), they are equal, *i.e.*

$$\Pi_j = P_j. \tag{1.9}$$

This relation is called *PASTA* (Poisson arrivals see time average), and it results from the Markov property of the exponential distribution [11]. The term PASTA comes from the fact that P_j is equal to the average (expected) time fraction of j calls existing when observed over a sufficiently long period. It is worth noting that other processes besides Poisson can also see time average, which is called *ASTA* (arrivals see time average). Such processes may appear in queuing network problems [12].

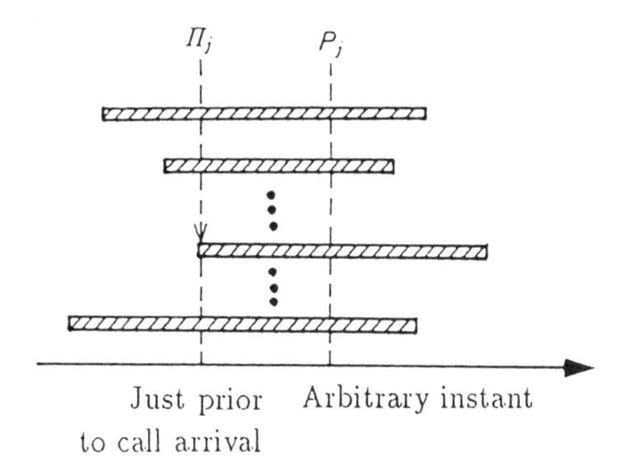

Fig.1.8 Probabilities of system state

If we aggregate n independent Poisson streams with rates λ_j, $j = 1, 2, \cdots, n$, as shown in Figure 1.9(a), then the resultant stream again becomes a Poisson stream with rate $\lambda = \lambda_1 + \lambda_2 + \cdots + \lambda_n$. This is because the convolution of Poisson distributions is again Poissonian. (See Appendix C.4.2.) If a Poisson stream with rate λ is directed to route j with probability p_j as in Figure 1.9(b), the stream in route j again becomes Poissonian. These properties are useful for analysing systems with Poisson input.

Fig.1.9 Aggregation and decomposition of Poisson stream

1.3.3 Little Formula

For a system in steady state, let λ be the arrival rate, L the mean number of waiting calls, and W the mean waiting time, as shown in Figure 1.10. Then, we have in general the relation,

$$L = \lambda W \tag{1.10}$$

which is called the *Little formula.*

Equation (1.10) is interpreted as follows: Since W is the mean sojourn time calls in the queue, if it is regarded as the mean holding time of the waiting calls, the right hand side of (1.10) is the traffic load of the waiting call, which is equal to the mean number of waiting calls from Property (4) in Subsection 1.2.1. It should be noted that L is the average value seen by outside observers, while W is one experienced by arrivals.

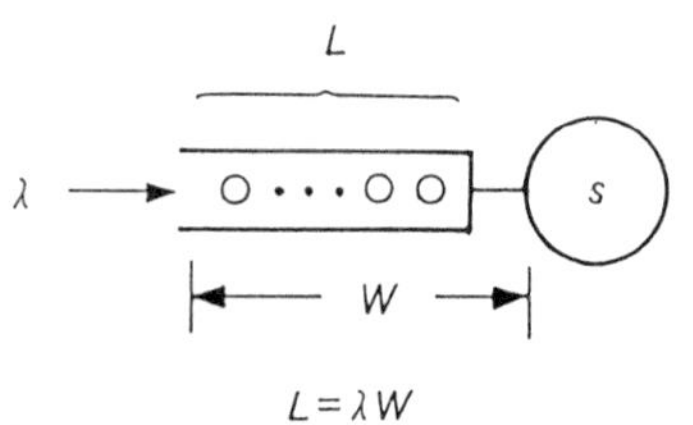

Fig.1.10 Application of Little formula

The *system time* is defined as the total sojourn time (waiting time + service time) spent by a call. Letting T be the mean system time, and N the mean number of calls existing (waiting and served)in the system, we obtain a variant of the Little formula,

$$N = \lambda T. \tag{1.11}$$

The Little formula is applicable to general G/G/s systems regardless of input process, service mechanism and queue discipline.

[Example 1.2] Suppose that a bank is visited by 24 customers per hour on average, and they form a queue when all tellers are busy, as shown in Figure 1.11. Since the arrival rate is $\lambda = 24/60 = 0.4/\text{min}$, if the mean number of waiting customers $L = 2.4$ is observed, from (1.10) the mean waiting time W for customers is given by

$$W = L/\lambda = 2.4/0.4 = 6\,\text{min}.$$

If the mean service time is $h = 5\,\text{min}$, the offered traffic load is $a = \lambda h = 2\,\text{erl}$ and thus the mean number of customers being served is 2 from Property (4) in Subsection 1.2.1. Hence, the mean number of customers existing in the bank is $N = L + a = 4.4$. This may be calculated alternatively as follows: Since the mean system time (staying time in the bank) is $T = W + h = 11\,\text{min}$, we have from (1.11)

$$N = \lambda T = 0.4 \times 11 = 4.4.$$

Since one teller can serve 1 erl at maximum, if the number of tellers is $s \leq 2$, the queue builds up to infinity and no steady state exists. In fact, it is known that a steady state exists if and only if $s > a$. (See Subsection 2.2.1.)

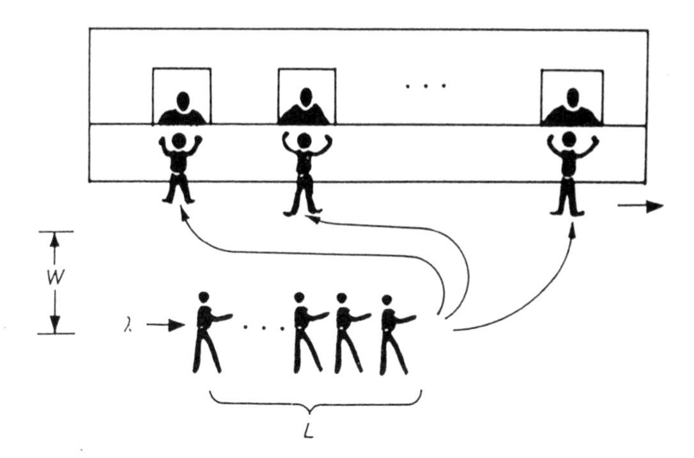

Fig.1.11 Example of queueing model

Exercises

[1] Suppose that a trunk group with 5 circuits carries 50 calls per hour, and the mean service time is 3 min. Calculate

(1) The traffic load carried by one circuit.

(2) The mean number of busy circuits.

[2] When 10 calls on average arrive per hour at random, find

(1) The probability that 2 or more calls originate in 12 min.

(2) The probability that interarrival time is no greater than 6 min.

[3] Assuming that the service time is exponentially distributed with mean 3 min, estimate

(1) The probability that the service time exceeds 6 min.

(2) The mean time until a call terminates when 6 calls are in progress.

[4] In a coin telephone station with more than 2 telephones, 50 customers per hour visit to make calls for 3 min on average. Calculate

(1) The mean number of telephones being used.

(2) The mean waiting time, when 1.2 waiting customers are observed on average.

Chapter 2

MARKOVIAN MODELS

Markovian models with interarrival and service times both exponentially distributed have been studied for a long time along with the development of telephone exchanges, and they have been widely applied in that field. This chapter presents a traditional analysis and the main results for the Markovian models.

2.1 Markovian Loss Systems

2.1.1 M/M/s(0)

Consider a full availability loss system with Poisson arrival and exponential service time as shown in Figure 2.1. Since arrivals finding all servers busy are immediately cleared, the number of calls present in the system is the same as those being served, *i.e.* the number of busy servers.

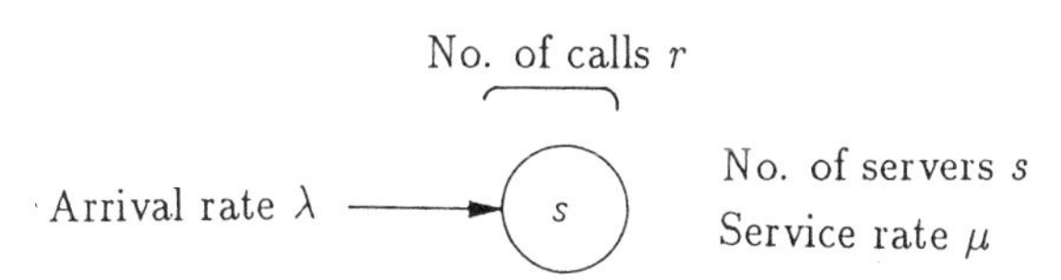

Fig.2.1 M/M/s(0) model

Let $N(t)$ be the number of calls existing at time t, and neglect the events that two or more calls originate or terminate in $(t, t + \Delta t]$ as $\Delta t \to 0$. Then, the event $\{N(t + \Delta t) = r\}$ results from one of the following 3 events:

A : $N(t) = r$, and no calls originate or terminate in $(t, t + \Delta t]$.

B : $N(t) = r - 1$, and one call originates in $(t, t + \Delta t]$.

C : $N(t) = r + 1$, and one call terminates in $(t, t + \Delta t]$.

Denote by λ and μ the arrival rate and service rate, respectively. Noting that the probability of one call terminating in $(t, t+\Delta t]$ with r calls in progress is $r\mu\Delta t$, and letting $P_r(t) = P\{N(t) = r\}$, we have the probabilities of the above events,

$$\begin{aligned} P\{A\} &= P_r(t)(1 - \lambda\Delta t - r\mu\Delta t) \\ P\{B\} &= P_{r-1}(t)\lambda\Delta t \\ P\{C\} &= P_{r+1}(t)(r+1)\mu\Delta t. \end{aligned} \tag{2.1}$$

Since these events are exclusive, we have (See Appendix C [T1].)

$$\begin{aligned} P_r(t+\Delta t) &= P\{A\} + P\{B\} + P\{C\} \\ &= P_r(t) + [\lambda P_{r-1}(t) - (\lambda + r\mu)P_r(t) + (r+1)\mu P_{r+1}(t)]\Delta t \end{aligned} \tag{2.2}$$

which yields the difference-differential equation,

$$\lim_{\Delta t \to 0} \frac{P_r(t+\Delta t) - P_r(t)}{\Delta t} =$$

$$\frac{dP_r(t)}{dt} = \lambda P_{r-1}(t) - (\lambda + r\mu)P_r(t) + (r+1)\mu P_{r+1}(t). \tag{2.3}$$

In practice, we are interested in the steady state when a long time has elapsed from the beginning. It is known that if the steady state exists, there is a unique limiting probability distribution $\{P_r\}$ such that for $t \to \infty$

$$P_r(t) \to P_r, \quad \frac{dP_r(t)}{dt} \to 0$$

independent of the initial condition. (See Appendix C.5.3.) This is called the *statistical equilibrium*, and P_r is referred to as the *steady state (equilibrium) probability*. Hence, in the steady state, the left hand side of (2.3) vanishes, and we have the difference equation,

$$(\lambda + r\mu)P_r = \lambda P_{r-1} + (r+1)\mu P_{r+1}, \quad r = 0, 1, \cdots, s \tag{2.4}$$

where $P_r = 0$ for $r = -1, s+1$.

To solve the difference equation, summing side by side of (2.4) from $r = 0$ to $i-1$, and setting $a = \lambda/\mu$, we have the *recurrence formula*,

$$P_i = \frac{a}{i}P_{i-1}, \quad i = 1, 2, \cdots, s. \tag{2.5}$$

Successive applications of (2.5) yields

$$P_i = \frac{a}{i}P_{i-1} = \frac{a^2}{i(i-1)}P_{i-2} = \cdots = \frac{a^i}{i!}P_0 \tag{2.6}$$

where P_0 is the probability that the system is empty, *i.e.* no calls exist in the system. From the *normalization condition*,

$$\sum_{i=0}^{s} P_i = P_0 + P_0 \sum_{i=1}^{s} \frac{a^i}{i!} = 1, \tag{2.7}$$

it follows that

$$P_0 = \left(\sum_{i=0}^{s} \frac{a^i}{i!} \right)^{-1}. \tag{2.8}$$

Using (2.8) in (2.6), we get the *Erlang distribution*,

$$P_r = \frac{\dfrac{a^r}{r!}}{\displaystyle\sum_{i=0}^{s} \frac{a^i}{i!}}, \quad r = 0, 1, \cdots, s. \tag{2.9}$$

In (2.9), letting s become large, we have

$$P_r \to \frac{a^r}{r!} e^{-a}, \quad s \to \infty \tag{2.9a}$$

which is no other than the Poisson distribution. (See Fig.2.4.) Thus, (2.9) is interpreted as the Poisson input is truncated by the capacity of the outlets. For this reason, (2.9) is sometimes called the *truncated Poisson distribution*.

2.1.2 Statistical Equilibrium

Now, let us consider the physical meaning of the statistical equilibrium. If both sides of (2.4) are multiplied by Δt, we obtain

$$P_r \lambda \Delta t + P_r r \mu \Delta t = P_{r-1} \lambda \Delta t + P_{r+1}(r+1)\mu \Delta t \tag{2.10}$$

which is interpreted as follows:

Denote the state of r calls present in the system by S_r. Then the left hand side of (2.10) represents the probability that $S_r \to S_{r+1}$ or $S_r \to S_{r-1}$, while the right hand side represents the probability that $S_{r-1} \to S_r$ or $S_{r+1} \to S_r$. Hence, (2.10) implies that the probability outgoing from S_r to the neighbor states, is equal to the probability incoming from the neighbor states to S_r. This is called *rate-out=rate-in* and expressed in the *state transition diagram* as shown in Figure 2.2, and (2.4) is called the (*equilibrium*) *state equation*. It will be suggested that if we are interested only in the steady state which is assumed to exist, we can obtain the state equation directly from the state transition diagram, and then to calculate the steady state probability, without the argument through (2.1) to (2.3) above.

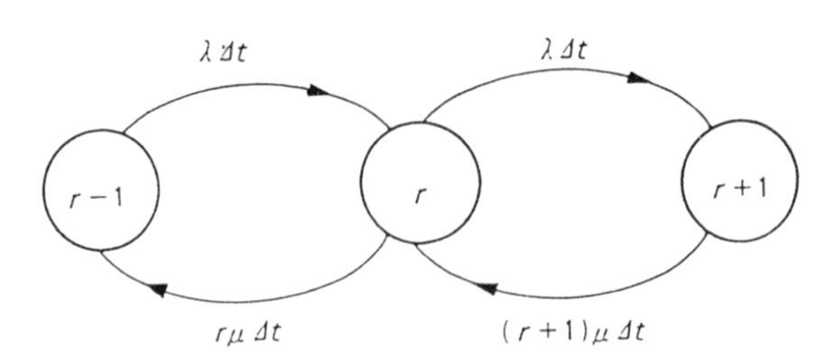

Fig.2.2 State transition diagram of M/M/$s(0)$

In this model, from (2.5) we have also the relation,

$$\lambda P_{r-1} = r\mu P_r, \quad r = 1, 2, \cdots, s. \tag{2.10a}$$

The left hand side means the rate for the state to go up, and the right hand side to go down, and thus (2.10a) is called *rate-up=rate-down*. The condition of (2.10a) is also referred to as *local balance*, while that of (2.4) *global balance*. It should be noted that the global balance and the local balance are not vice versa in general, *e.g.* even if the global balance holds, the local balance does not necessarily hold in general.

It is suggested that if the steady state exists, we can obtain the state equations (2.4) or (2.10a) directly from the state transition diagram, without the argument (2.1) to (2.3) above. Although the existence of steady state is not necessarily guaranteed in general, it is known that it always exists in loss systems.

In the sequel, we shall use these relations to analyse systems in the steady state.

2.1.3 M(n)/M/s(0)

Consider a similar loss system as before except for finite sources of n inlets. Assume that the interarrival time from an idle inlet is exponentially distributed with mean ν^{-1}. This is called *quasi-random input (process, arrival, etc.)*. Using the similar notation as before, we shall analyse the system.

If the r calls exist in the system, $(n-r)$ inlets are idle, and the effective arrival rate is $(n-r)\nu$. Hence, we have the state transition diagram as shown in Figure 2.3. From the rate-out=rate-in, we get the steady state equation,

$$\begin{aligned} [(n-r)\nu + r\mu]P_r &= (n-r+1)\nu P_{r-1} + (r+1)\mu P_{r+1}, \\ & \qquad r = 0, 1, \cdots, s\,;\; P_{-1} = P_{s+1} = 0. \end{aligned} \tag{2.11}$$

From a similar procedure to (2.5), we have the recurrence formula,

$$P_r = \frac{(n-r+1)\nu h}{r} P_{r-1}, \quad r = 1, 2, \cdots, s \tag{2.11a}$$

where $h = \mu^{-1}$ is the mean service time. Equation (2.11a) could have been obtained directly from the rate-up = rate-down by inspecting Figure 2.3. Using (2.11a) successively and the normalization condition, yield the *Engset distribution*,

$$P_r = \frac{\binom{n}{r}(\nu h)^r}{\sum_{i=0}^{s}\binom{n}{i}(\nu h)^i}, \quad r = 0, 1, \cdots, s. \tag{2.12}$$

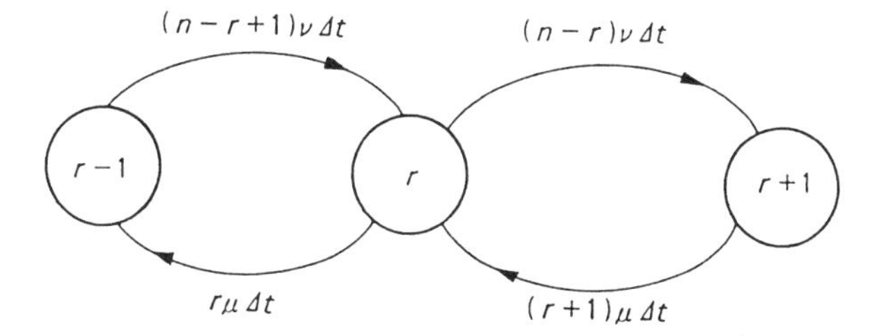

Fig.2.3 State transition diagram of M(n)/M/s(0)

Letting $n \to \infty$ keeping $n\nu h = a$ fixed, we have

$$\binom{n}{r}(\nu h)^r = \frac{n(n-1)(n-2)\cdots(n-r+1)}{n^r}\frac{(n\nu h)^r}{r!} \to \frac{a^r}{r!}. \tag{2.13}$$

Then, (2.12) coincides with the Erlang distribution in (2.9) with the offered load a. Therefore, M/M/s(0) is interpreted as the corresponding infinite source model.

On the other hand, if $n \leq s$, since the denominator of (2.12) becomes $(1+\nu h)^n$, (2.12) reduces to the *binomial distribution*,

$$P_r = \binom{n}{r}\alpha^r(1-\alpha)^{n-r} \tag{2.14}$$

where

$$\alpha = \frac{\nu h}{1+\nu h}. \tag{2.14a}$$

Thus, (2.12) provides a unified formula giving the steady state probability of loss systems. Figure 2.4 shows numerical examples. It will be seen that in the order of Poisson→Erlang→Engset→Binomial, the probability distribution becomes narrow about the mean, *i.e.* the variance tends to decrease.

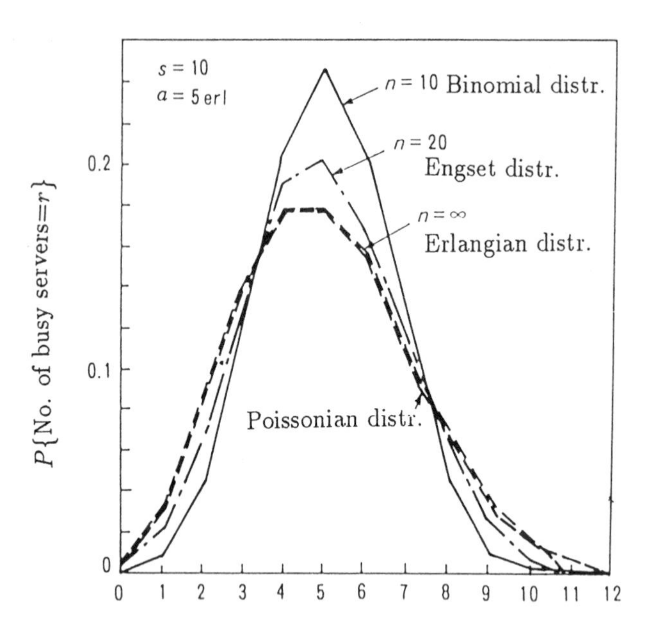

Fig.2.4 Comparison of various distributions

2.1.4 State Probabilities at Call Arrival

Let us find the probability Π_r that r calls exist in the system just prior to a call arrival in the steady state. Denote the event that a call arrives in Δt by ΔE, and that of r customers existing in the steady state by E_r. Then, since Π_r is the conditional probability of E_r, given ΔE, we have from *Bayes' theorem*, (See Appendix C [T2].)

$$\Pi_r = P\{E_r \mid \Delta E\} = \frac{P\{E_r\}P\{\Delta E \mid E_r\}}{\sum_{i=0}^{s} P\{E_i\}P\{\Delta E \mid E_i\}}. \tag{2.15}$$

Noting that $P\{\Delta E \mid E_r\} = (n-r)\nu\Delta t$, and $P\{E_r\} = P_r$ given by (2.12), we have

$$\begin{aligned} P\{E_r\}P\{\Delta E \mid E_r\} &= \binom{n}{r}(\nu h)^r P_0(n-r)\nu\Delta t \\ &= \binom{n-1}{r}(\nu h)^r P_0 n\nu\Delta t. \end{aligned}$$

Using this in (2.15), we obtain

$$\Pi_r = \frac{\binom{n-1}{r}(\nu h)^r}{\sum_{i=0}^{s}\binom{n-1}{i}(\nu h)^i}, \quad r = 0, 1, \cdots, s. \tag{2.16}$$

If (2.16) is compared with (2.12), it can be seen that Π_r with n inlets is equivalent to P_r with $(n-1)$ inlets, and this relation is expressed by

$$\Pi_r[n] = P_r[n-1]. \tag{2.17}$$

This may be interpreted as follows: Since an arriving customer occupies one inlet, he will see the behavior of the system with the remaining $(n-1)$ inlets. If we let $n \to \infty$, the input becomes Poissonian and we have

$$\Pi_r[\infty] = P_r[\infty] \tag{2.18}$$

which would have been expected from the PASTA.

2.1.5 Blocking Probability

Denote the offered traffic load to, and carried load by a system, respectively, by a, and a_c. Then, the *loss rate* is defined by

$$B = \frac{a - a_c}{a} \tag{2.19}$$

as the measure of GOS, which is often called the *blocking probability.*

Letting N be the number of calls existing in the steady state, from Property (4) in Subsection 1.2.1, the carried load a_c is the expectation of N. Hence, we have

$$a_c = E\{N\} = \sum_{r=0}^{s} rP_r = P_0 n\nu h \sum_{r=0}^{s-1}\binom{n-1}{r}(\nu h)^r. \tag{2.20}$$

The offered load a is evaluated from (1.1) as

$$a = E\{N-n\}\nu h = P_0 n\nu h \sum_{r=0}^{s}\binom{n-1}{r}(\nu h)^r. \tag{2.21}$$

substituting (2.20) and (2.21) into (2.19), we obtain the *Engset loss formula,*

$$B = \frac{\binom{n-1}{s}(\nu h)^r}{\sum_{i=0}^{s}\binom{n-1}{i}(\nu h)^i} = \Pi_s. \tag{2.22}$$

This is equivalent to Π_s given by (2.16), the probability that an arrival finds all servers busy. Thus, the loss rate is equivalent to the probability that a call is blocked, and (2.22) is also called the *call congestion probability*.

On the other hand, the probability that an outside observer finds all servers busy is obtained from (2.12) as

$$B_T = \frac{\binom{n}{s} (\nu h)^r}{\sum_{i=0}^{s} \binom{n}{i} (\nu h)^i} = P_s. \tag{2.23}$$

This is called the *time congestion probability* since it tends to the time fraction of all servers busy over a sufficiently long time observation.

We can derive the relation between a and B as

$$a = \frac{n\nu h}{1 + \nu h(1 - B)}. \tag{2.24}$$

It should be noted that even if n, ν and h are given, a is dependent on B in general. However, since $B = 0$ in the case of $n \leq s$ where whole offered load is carried, we have

$$a = \frac{n\nu h}{1 + \nu h} \equiv a_S \tag{2.25}$$

where a_S is called the *intended traffic load*. In this case, the state probability P_j reduces to the binomial distribution as shown in (2.14), and (2.14a) becomes $\alpha = a_S/n$ which is the offered load per inlet.

Letting $n \to \infty$ and $\nu \to 0$ while keeping $n\nu h = a$ fixed, and using the relation (2.13), we obtain the *Erlang loss formula*,

$$B = B_T = \frac{\frac{a^s}{s!}}{\sum_{i=1}^{s} \frac{a^i}{i!}} \equiv E_s(a) \tag{2.26}$$

which is also referred to as the *Erlang B formula* and denoted by $E_s(a)$. In this case, we have the relation that the offered load is equal to the intended load,

$$a = \lim_{n \to \infty} \frac{n\nu h}{1 + \nu h(1 - B)} = a_S. \tag{2.27}$$

Because (2.26) has a simplicity that the offered load is independent of B, and it gives a pessimistic estimation (actual GOS is better than estimated) as seen in Figure 2.5, the Erlang loss formula has been widely used in telephone applications.

From (2.26), we have the recurrence formula,

$$E_s(a) = \frac{aE_{s-1}(a)}{s + aE_{s-1}(a)}, \quad E_0(a) = 1 \tag{2.26a}$$

which is convenient for numerical computation. The Erlang B formula is useful for teletraffic engineering, and relevant computer programs are given in Appendix B.

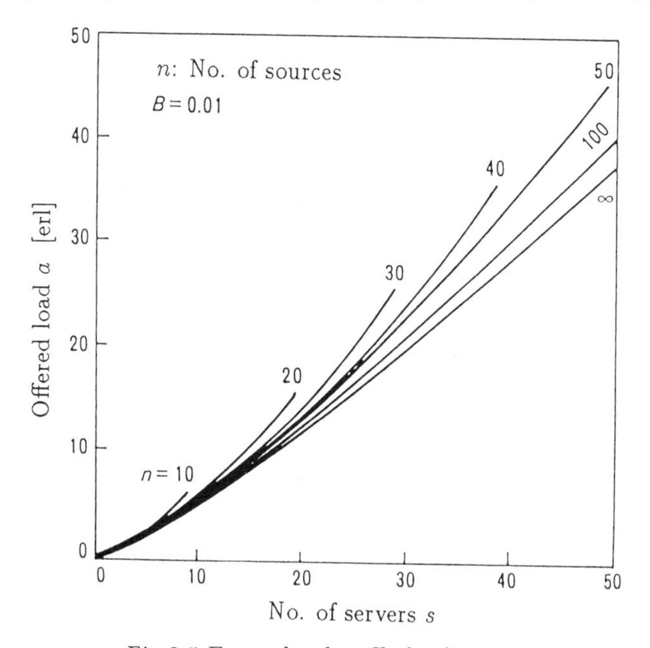

Fig.2.5 Example of traffic load curves

In practice, it is often necessary to determine the number of trunks required for carrying a given traffic load at a given GOS. Curves indicating the relation between the offered load a and the number of trunks s providing a given GOS, are called *traffic load curves*, for which an example is shown in Figure 2.5. The curve with $n = \infty$ corresponds to the Erlang B formula. It will be seen that the Erlang B formula gives a pessimistic estimation as stated.

The traffic load carried by a trunk is called the *trunk efficiency* or *occupancy* and is given by

$$\eta = \frac{a_c}{s} = \frac{a(1-B)}{s}. \tag{2.28}$$

It should be noted that since a trunk can carry 1 erl at maximum, we have the relation, $\eta \leq 1$. Hence, in general for loss systems, even if the offered load $a > s$, the carried load is $a_c = a(1-B) < s$, and thus a steady state always exists as stated before.

Figure 2.6 shows a numerical example of the trunk efficiency, given the blocking probability $B = 0.01$. It will be seen that with a given GOS (blocking probability), the greater the number of trunks the higher becomes the trunk efficiency. This is called the *large scale effect*, and it is the case in a general traffic system, which is an important feature in designing telecommunications and computer systems.

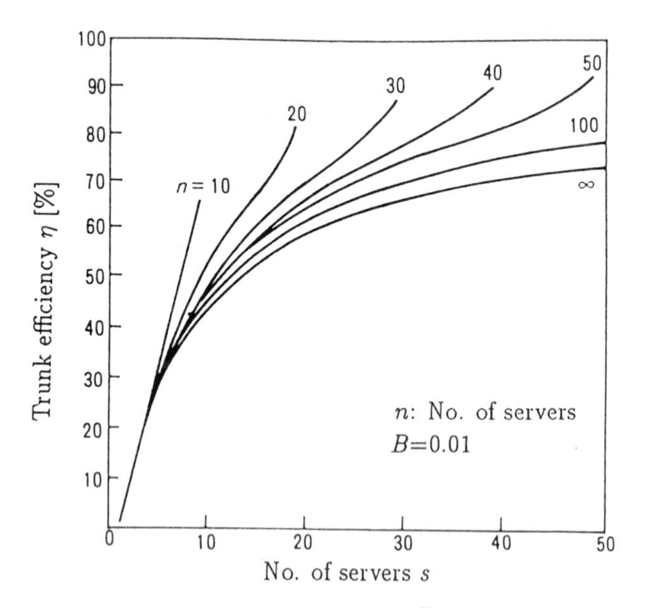

Fig.2.6 Example of trunk efficiency

[Example 2.1] Let us design a trunk group between two PBXs with the blocking probability $B \leq 0.01$, as in Figure 1.1.

In general, a telephone is used either as *calling party* or *called party*. The *calling rate* is defined by the traffic load per telephone used as calling party. In this example, since a telephone is used once an hour for 6 min, the traffic load per telephone is 0.1 erl, of which 50 % are assumed to be origination calls on average, and thus the calling rate becomes 0.05 erl.

First, we shall design a one-way trunk group as shown in Figure 2.7(a). Assuming that the connection is made uniformly within the two PBXs, the traffic load from A to B is

$$a_A = 0.05 \times 1000 \times \frac{1}{2} = 25\,\text{erl}.$$

Since the number of sources (telephones) is sufficiently large and telephones are used at random, applying M/M/s(0) model, we can calculate the blocking probability by Erlang B formula. Using Program 1 in Appendix B, we obtain

$$E_{35}(25) = 0.0116 > 0.01, \quad E_{36}(25) = 0.0080 < 0.01.$$

Hence, the required number of trunks from A to B is $s_A = 36$. Similarly, from B to A, we have the traffic load $a_B = 25$ erl and the number of trunks $s_B = 36$, thus the total number of trunks required is $s = s_A + s_B = 72$. In this case, from (2.28) the trunk efficiency is

$$\eta = \frac{25(1 - 0.0080)}{36} = 68.9\,\%.$$

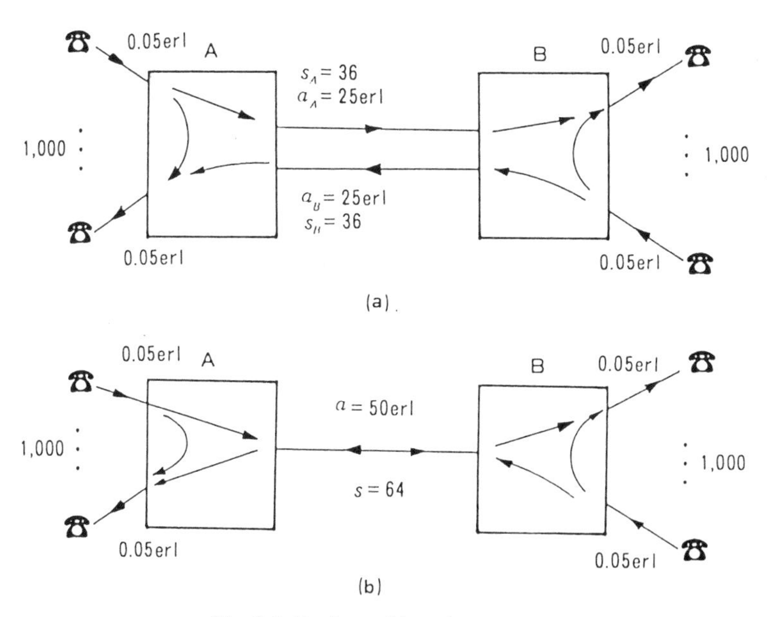

Fig.2.7 Design of trunk group

Next, we shall try to use a two-way trunk group which carries the total traffic between A and B as shown in Figure 2.7(b). With the two-way traffic load,

$$a = a_A + a_B = 50\,\text{erl}$$

similarly we can calculate $s = 64$ with $E_{64}(50) = 0.0084 < 0.01$, and the trunk efficiency,

$$\eta = \frac{50(1 - 0.0084)}{64} = 77.5\,\%.$$

It can be seen that the efficiency is increased by virtue of the large scale effect. It should be noted, however, that in the two-way operation some additional facilities are needed for preventing collisions or double connections with simultaneous usage of both ends. Therefore, the optimum design is required including such an additional expenditure [5, p.159].

In telephone practice, the traffic statistics for the busiest hour, say 10:00 to 11:00 AM, averaged over a year, is used for traffic design, and over-loaded conditions are taken into account, *e.g.* for the 10 high days, such as Christmas, Thanksgiving, accidental events, etc. Supposing an over-loaded condition in which the calling rate is increased by 50%, we have

One-way system: $a_A = a_B = 25 \times 1.5 = 37.5\,\text{erl}$, $E_{36}(37.5) = 0.1444$

Two-way system: $a = 50 \times 1.5 = 75\,\text{erl}$, $E_{64}(75) = 0.1924$.

It will be recognized that although the blocking probability is increased by one order, the loss system is still stable for the overload condition, and the one-way system provides a better GOS than the two-way system.

2.1.6 Ordered Trunk Hunting

Consider the ordered hunting model shown in Figure 2.8, in which the traffic load a erl is offered to enumerated n trunks, which are hunted from the smaller to larger number. The traffic load carried by the rth trunk is given by

$$a_r = a[E_{r-1}(a) - E_r(a)] \tag{2.29}$$

for which a numerical example is shown in Figure 2.9. It will be seen that the larger the number the less traffic load is carried. In particular, the traffic load carried by the last (sth) trunk is called the *last trunk capacity* (LTC).

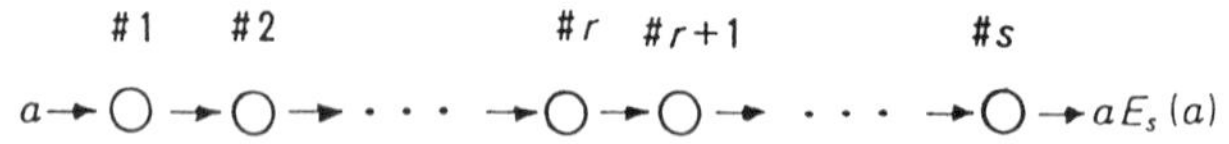

Fig.2.8 Model of ordered hunting

The total carried load a_c is clearly

$$a_c = \sum_{r=1}^{s} a_r = a[1 - E_s(a)] \tag{2.30}$$

which is independent of the way of hunting, *e.g.* sequential, random, etc., since no such assumptions were made in deriving $E_s(a)$ in (2.26).

The *additional trunk capacity* (ATC) is defined by the traffic load increment Δa when one trunk is added while the blocking probability B is fixed. Hence, we have

$$E_s(a) = E_{s+1}(a + \Delta a) = B. \tag{2.31}$$

The LTC and ATC are used in an optimum design of alternative routing systems. (See Subsection 5.3.1.)

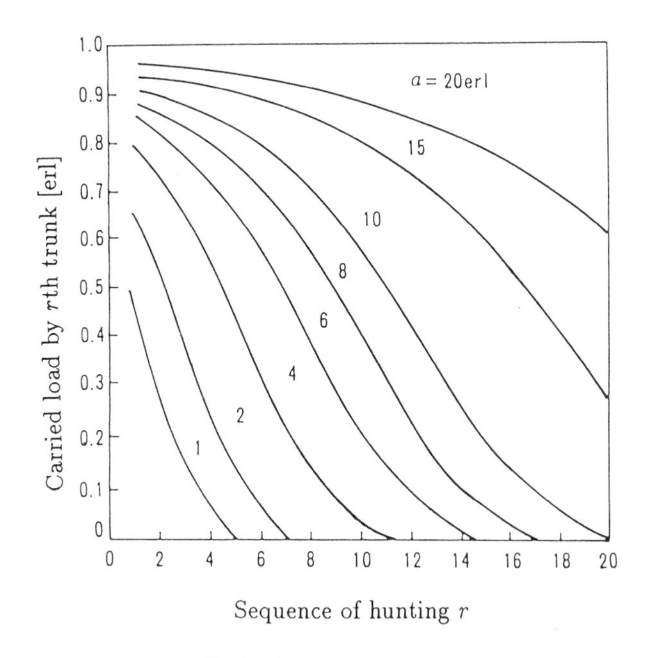

Fig.2.9 Traffic load carried by ordered hunting

2.2 Markovian Delay Systems

2.2.1 M/M/s

Consider the delay system M/M/s with Poisson input, exponential service time, s servers, and infinite waiting room as shown in Figure 2.10, in which all waiting calls wait until they become served.

Letting λ and μ be the arrival rate and service rate, respectively, we have the state transition diagram shown in Figure 2.11. The diagram for $r < s$ is equivalent to that of M/M/$s(0)$ in Figure 2.2. On the other hand, in the diagram for $r \geq s$, since s calls are in service and the remaining $(r-s)$ calls are waiting, the probability that a call terminates in Δt, is $s\mu\Delta t$ regardless of r.

If a steady state exists, from the rate-out=rate-in, we have the steady state equations,

$$\begin{aligned}(\lambda + r\mu)P_r(t) &= \lambda P_{r-1}(t) + (r+1)\mu P_{r+1}(t), \quad r < s \\ (\lambda + s\mu)P_r(t) &= \lambda P_{r-1}(t) + s\mu P_{r+1}(t), \qquad r \geq s.\end{aligned} \tag{2.32}$$

Solving (2.32) similarly as (2.4), and letting $a = \lambda/\mu$, we get

$$\begin{aligned}P_r &= \frac{a^r}{r!}P_0, \qquad r < s \\ P_r &= \frac{a^s}{s!}\left(\frac{a}{s}\right)^{r-s} P_0, \quad r \geq s.\end{aligned} \tag{2.33}$$

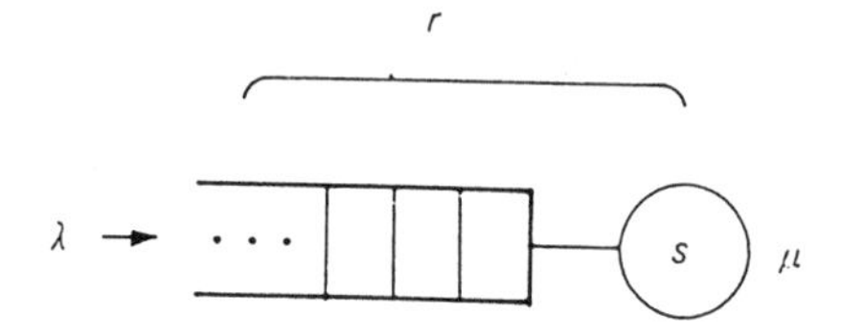

Fig.2.10 M/M/s model

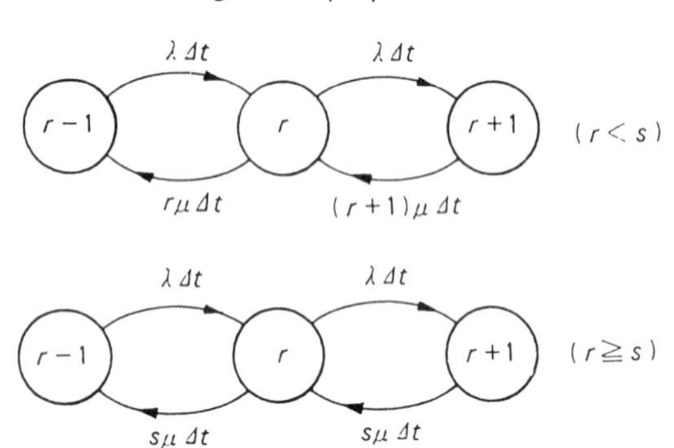

Fig.2.11 State transition of M/M/s

If $\{P_r\}$ is the equilibrium probability distribution, the following normalizing condition should hold:

$$\sum_{r=0}^{\infty} P_r = P_0 \left[\sum_{r=0}^{s-1} \frac{a^r}{r!} + \frac{a^s}{s!} \sum_{r=0}^{\infty} \left(\frac{a}{s}\right)^r\right] = 1.$$

If $a < s$, the second summation in [ ] converges as

$$\sum_{r=0}^{\infty} \left(\frac{a}{s}\right)^r = \frac{s}{s-a}$$

and we get

$$P_0 = \left(\sum_{r=0}^{s-1} \frac{a^r}{r!} + \frac{s}{s-a}\right)^{-1}. \tag{2.34}$$

If $a \geq s$, the summation diverges and we have $P_0 = 0$, which means that $\{P_r\}$ does not exist. In fact, since no calls are lost in a delay system, the offered load a erl has to be carried. However, since s trunks can carry only s erl, if $a \geq s$, the waiting calls build up to infinity and the system diverges, and hence no steady state exists. The *utilization factor* is defined by $\rho = a/s$, which is the carried load per trunk (server). It is known in general that a delay system has a steady state if and only if $\rho < 1$. This condition should be kept in mind in designing delay systems, and it is assumed unless stated otherwise in the sequel.

2.2.2 Mean Waiting Time

The *waiting probability* is defined as the probability that an arrival has to wait, and is denoted by $M(0)$ which means the probability that the waiting time is greater than 0. An incoming call has to wait if and only if the number of calls present in the system is no smaller than s. Hence, noting the PASTA, we have

$$M(0) = \sum_{r=s}^{\infty} P_r = \frac{a^s}{s!}\frac{s}{s-a}P_0 = \frac{\dfrac{a^s}{s!}\dfrac{s}{s-a}}{\displaystyle\sum_{r=0}^{s-1}\frac{a^r}{r!} + \frac{s}{s-a}} \tag{2.35}$$

which is called the *Erlang C formula*. Rearranging (2.35), we have

$$M(0) = \frac{sE_s(a)}{s - a[1 - E_s(a)]} \tag{2.36}$$

which is convenient for calculation using the Erlang B formula.

The mean number of waiting calls is given by

$$\begin{aligned} L &= \sum_{r=s}^{\infty}(r-s)P_r = \frac{a^s}{s!}P_0\sum_{r=0}^{\infty} r\left(\frac{a}{s}\right)^r \\ &= M(0)\frac{a}{s-a} \end{aligned} \tag{2.37}$$

where we used the identity,

$$\sum_{r=0}^{\infty} r x^r = \frac{x}{(1-x)^2}, \quad x < 1.$$

From the Little formula (1.10), we have the mean waiting time,

$$W = \frac{L}{\lambda} = M(0)\frac{h}{s-a} \tag{2.38}$$

where $h = \mu^{-1}$ is the mean service time, and $a = \lambda/\mu$ the offered traffic load. Figure 2.12 shows a numerical example.

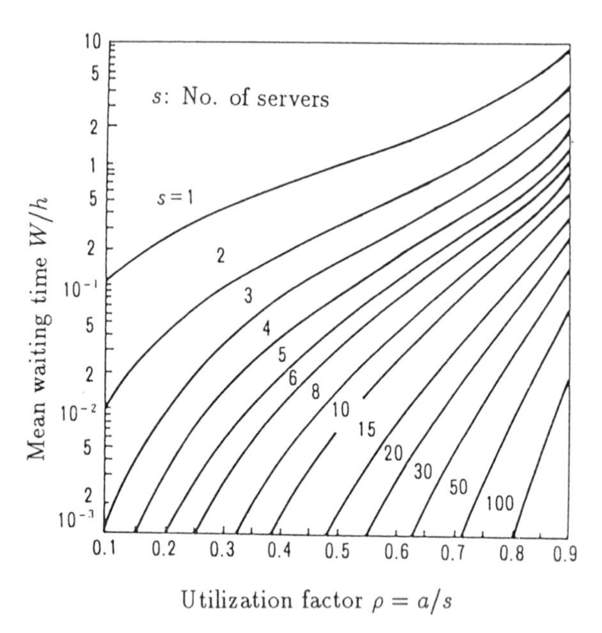

Fig.2.12 Mean waiting time of M/M/s

2.2.3 Waiting Time Distribution

In evaluating the waiting time distribution, the queue discipline should be specified for serving the waiting calls. First, let us assume the FIFO discipline.

Marking an arbitrary call, it is referred to as *test call.* The conditional probability q_j that j calls are waiting, given the test call has to wait, is given by

$$q_j = \frac{P_{s+j}}{M(0)} = (1-\rho)\rho^j, \quad j = 0, 1, \cdots \tag{2.39}$$

which follows a *geometric distribution.* Taking the origin at the test call arrival epoch, since the service time is exponentially distributed and s calls are in service, the probability that a call is terminated in time $(t, t+\Delta t]$ is $s\mu\Delta t$. Hence, from a similar argument as in (1.2), the probability that k calls terminate in time $(0, t]$ becomes a Poisson distribution with mean $s\mu t$. Since the test call enters service when j calls waiting before it have been served, the conditional probability $Q_j(t)$ that the test call has to wait longer than t, given j calls are waiting, is expressed by

$$Q_j(t) = e^{-s\mu t} \sum_{k=0}^{j} \frac{(s\mu t)^k}{k!}. \tag{2.40}$$

Since we selected the test call arbitrarily, the complementary waiting time distribution function $M(t)$ (probability of waiting time exceeding t) is given by, from the total probability theory (See Appendix C [T2].),

$$M(t) = \sum_{j=0}^{\infty} Q_j(t) q_j = (1-\rho)e^{-s\mu t} \sum_{k=0}^{j} \frac{(s\mu t)^k}{k!}. \tag{2.40a}$$

Using the identity,

$$\sum_{r=0}^{\infty} \sum_{k=0}^{r} f(r,k) = \sum_{r=0}^{\infty} \sum_{k=0}^{\infty} f(r+k,k)$$

and rearranging (2.40a), finally we have

$$M(t) = M(0)e^{-(1-\rho)st/h}. \tag{2.41}$$

A numerical example is shown in Figure 2.13.

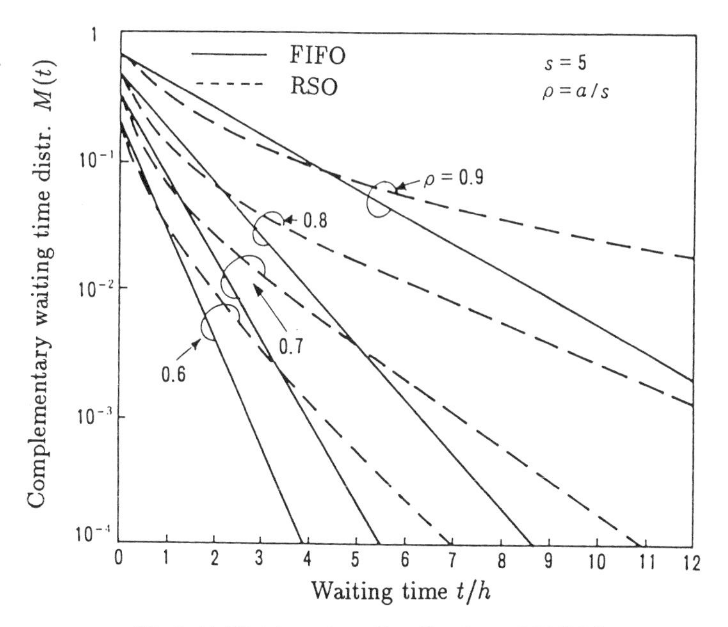

Fig.2.13 Waiting time distribution of M/M/s

Next, let us consider the model with RSO (random service order). Although the exact solution is not of simplified form, it is approximated by H_2 (hyper-exponential distribution) matching the first 3 moments, as [3, p.87]

$$M(t) = \frac{1}{2} M(0) \left[r_1 e^{-r_1(s-a)t/h} + r_2 e^{-r_2(s-a)t/h} \right] \tag{2.42}$$

where

$$r_1 = 1 - \sqrt{\frac{a}{2s}}, \quad r_2 = 1 + \sqrt{\frac{a}{2s}}.$$

A numerical example is also shown in Figure 2.13. It will be seen that the probability of waiting longer, is larger than that for FIFO.

The disciplines such as FIFO, RSO, LIFO, etc., in which waiting calls are served independently of their service time, are classified in the *non-biased discipline.* For a non-biased discipline, the mean waiting time is the same and given by (2.38). On the other hand, disciplines such as SSTF (*shortest service time first*) dependent on the service time, are classified in the *biased discipline*, for which (2.38) is no longer valid, because the service time distribution of calls being served, differs from the original exponential service time.

[Example 2.2] In telephony, the time from the off hook (taking hand-set up) to the reception of the dial-tone is called *dial-tone delay.* In standard practice, the probability of dial-tone delay exceeding 3 sec is specified to be no greater than 1%. The originating registers (dial digit receivers) are provided to satisfy this criterion.

Suppose that in a PBX accommodating 3000 telephones each originating one call per hour in average, and the dialing time is exponentially distributed with mean 12 sec. Assuming that a fixed 0.5 sec is needed for connecting the register in the switching operation, then the waiting time allowed for the register becomes 2.5 sec.

The traffic load offered to the registers is

$$a = 3000 \times \frac{12}{3600} = 10 \,\text{erl}.$$

When the number of registers is s=17, from (2.36) we have $M(0) = 0.0309$. For the complementary waiting time distribution, from (2.41) and (2.42) we obtain

$$\text{FIFO}: \quad M(2.5\,\text{sec}) = 0.0072$$

$$\text{RSO}: \quad M(2.5\,\text{sec}) = 0.0061$$

both satisfy the requirement. The mean waiting time in both cases is $W = 0.0529$ sec from (2.38), thus the mean dial-tone delay is $T = 0.5529$ sec. (Check for $s = 16$.)

In the case of 100 % over-load, with $\rho = 20/17 > 1$, the steady state no longer exists and the dial-tone delay becomes infinite. Therefore, in practice, the utilization factor is limited to a certain value, say $\rho \leq 0.7$ under normal conditions so that it does not exceed unity under over-load conditions.

2.3 Extended Markovian Models

2.3.1 Birth-Death Process

A set of random variables $\{N(t); t \geq 0\}$ with parameter t, say the number of calls present in the system at time t, is called the *stochastic process*. The parameter t is often regarded as time, and if $N(t) = j$, the process is said to be in *state* j at time t. (See Appendix C.5.) In particular, if a stochastic process has the Markov property, *i.e.* the state after time t is dependent only on the state at time t, and independent of the progress before t, it is called the *Markov process*. (See Subsection 1.3.1.)

A Markov process, in which the state transition occurs only one step at a time, is called a *birth-death (B-D) process*, which is described by the relation, for $\Delta t \to 0$,

$$P\{N(t+\Delta t) = k \mid N(t) = j\} = \begin{cases} \lambda_j \Delta t, & k = j+1 \\ \mu_j \Delta t, & k = j-1 \\ 0, & \mid k-j \mid \geq 2\,;\ j,k = 0,1,\cdots. \end{cases} \tag{2.43}$$

where λ_j is called the *birth rate*, and μ_j the *death rate*, with $\mu_0 = 0$. (See Appendix C.5.4.) The state transition diagram is shown in Figure 2.14.

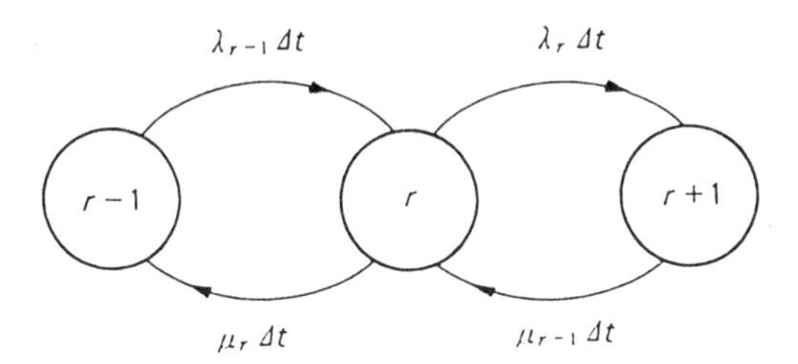

Fig.2.14 State transition for birth-death process

Assuming the existence of a steady state, and letting $P\{N(t) = r\} \to P_r$ as $t \to \infty$, from the rate-out=rate-in, we have the steady state equation,

$$\lambda_{r-1}P_{r-1} - (\lambda_r + \mu_r)P_r + \mu_{r+1}P_{r+1} = 0, \quad r = 0,1,\cdots;\ P_{-1} = 0. \tag{2.44}$$

Summing (2.44) over r similarly as in (2.4), and using the normalization condition, we have

$$\begin{aligned} P_r &= \frac{\lambda_0\lambda_1\cdots\lambda_{r-1}}{\mu_1\mu_2\cdots\mu_r} \\ P_0 &= \left(1 + \sum_{r=1}^{\infty} \frac{\lambda_0\lambda_1\cdots\lambda_{r-1}}{\mu_1\mu_2\cdots\mu_r}\right)^{-1}. \end{aligned} \tag{2.45}$$

It is known that a steady state exists if and only if $P_0 > 0$. The solutions for the Markovian models which appeared in the preceding sections all reduce to (2.45). We shall present below some examples for the extended models.

2.3.2 $\mathbf{M}(n)/\mathbf{M}/s(m,\xi)$

Consider the system with quasi-random input with n sources, exponential service time, and m waiting positions. Furthermore, it is assumed that the *early departure time* to give up waiting, is exponentially distributed with mean γ^{-1}. Letting ν be the arrival rate from an idle source, and μ^{-1} the mean service time, we have the birth rate and death rate, respectively, as

$$\begin{aligned} \lambda_j &= (n-j)\nu, & j &= 0,1,\cdots,s+m \\ \mu_j &= \begin{cases} j\mu, & j = 1,2,\cdots,s-1 \\ s\mu+(j-s)\gamma, & j = s,s+1,\cdots,s+m. \end{cases} \end{aligned} \tag{2.46}$$

Using (2.46) in (2.45), and noting the relation (2.17), we obtain

$$\begin{aligned} W &= \frac{h}{s}\Pi_0 \binom{n-1}{s} (\nu h)^s \sum_{r=0}^{m-1} \frac{(r+1)(n-1-s)_r}{\pi(r+1)} \left(\frac{\nu h}{s}\right)^r \\ B &= \Pi_0 \binom{n-1}{s} (\nu h)^s \frac{(n-1-s)_m}{\pi(m)} \left(\frac{\nu h}{s}\right)^s \\ M(0) &= \Pi_0 \binom{n-1}{s} (\nu h)^s \sum_{r=0}^{m-1} \frac{(n-1-s)_r}{\pi(r+1)} \left(\frac{\nu h}{s}\right)^r \\ M(t) &= \Pi_0 \binom{n-1}{s} \left(\frac{\nu h}{\xi}\right)^s \\ &\quad \times \sum_{r=0}^{m-1} \binom{n-1-s}{r} (\nu h)^r \sum_{i=0}^{r} (-1)^r \binom{r}{i} \frac{e^{-[s+(j+1)\xi]t/h}}{1+i\xi/s} \end{aligned} \tag{2.47}$$

where

$$\Pi_0 = \left[\sum_{r=0}^{s-1} \binom{n-1}{r} (\nu h)^r + \binom{n-1}{s} (\nu h)^s \sum_{r=0}^{m-1} \frac{(n-1-s)_r}{\pi(r)} \left(\frac{\nu h}{s}\right)^r\right]^{-1}$$

$$(n)_r = n(n-1)\cdots(n-r+1), \quad (n)_0 = 1; \quad (n)_r = 0, \quad r > n$$

$$\xi = \frac{\gamma}{\mu}, \quad \pi(r) = \prod_{i=0}^{r} \left(1 + i\frac{\xi}{s}\right).$$

The equations in (2.47) provide general formulas for Markovian models, from which performance evaluation formulas for various models may be derived as special cases. Some examples useful for application are shown below.

2.3.3 M/M/$s(m, \xi)$

In (2.47), letting $n \to \infty$ and $\nu \to 0$ with $n\nu h = a$, we obtain the formulas for Poisson input,

$$\begin{aligned}
W &= P_0 \frac{h}{s} \frac{a^s}{s!} \sum_{r=0}^{m-1} \frac{r+1}{\pi(r+1)} \left(\frac{a}{s}\right)^r \\
B &= P_0 \frac{a^s}{s!} \frac{1}{\pi(m)} \left(\frac{a}{s}\right)^m \\
M(0) &= P_0 \frac{a^s}{s!} \sum_{r=0}^{m-1} \frac{1}{\pi(r+1)} \left(\frac{a}{s}\right)^r \\
M(t) &= P_0 \frac{a^s}{s!} \sum_{r=0}^{m-1} \frac{a^r}{r!} \sum_{i=0}^{r} (-1)^i \binom{r}{i} \frac{e^{-[s+(j+1)\xi]t/h}}{1+i\xi/s}
\end{aligned} \tag{2.48}$$

where

$$P_0 = \left[\sum_{r=0}^{s-1} \frac{a^r}{r!} + \frac{a^s}{s!} \sum_{r=0}^{m-1} \frac{1}{\pi(r)} \left(\frac{a}{s}\right)^r\right]^{-1}.$$

With $m \to \infty$, letting $\xi \to \infty$ or $\xi \to 0$, $M(0)$ in (2.48) reduces to the Erlang B formula (2.26) or the C formula (2.35), respectively.

2.3.4 M(n)/M/s

Letting $m = n - s$ and $\xi \to \infty$ in (2.47), the model reduces to the finite source delay system, and we have

$$\begin{aligned}
W &= \frac{h}{s} \Pi_0 \binom{n-1}{s} (\nu h)^s \sum_{r=0}^{n-s-1} (r+1)(n-s-1)_r \left(\frac{\nu h}{s}\right)^r \\
M(0) &= \Pi_0 \binom{n-1}{s} (\nu h)^s \sum_{r=0}^{m-s-1} (n-s-1)_r \left(\frac{\nu h}{s}\right)^r \\
M(t) &= \Pi_0 \binom{n-1}{s} (\nu h)^s e^{-st/h} \sum_{r=0}^{n-s-1} (n-s-1)_r \left(\frac{\nu h}{s}\right)^r \sum_{i=0}^{r} \frac{(st/h)^i}{i!}
\end{aligned} \tag{2.49}$$

where

$$\Pi_0 = \left[\sum_{r=0}^{s-1} \binom{n-1}{r} (\nu h)^r + \binom{n-1}{s} (\nu h)^s \sum_{r=0}^{n-s-1} (n-s-1)_r \left(\frac{\nu h}{s}\right)^r\right]^{-1}.$$

A numerical example is shown in Figure 2.15. If we let $n \to \infty$, W in (2.49) coincides with that in (2.38).

Now, we shall derive a formula for calculating the offered load. Letting λ be the arrival rate, then the offered load is $a = \lambda h$ where h is the mean service time. Denoting the mean number of calls present in the system in steady state by $\overline{N}$, from the Little formula we have

$$\overline{N} = a + \lambda W.$$

Using this in

$$a = \nu h(n - \overline{N})$$

and solving it, we obtain

$$a = \frac{n\nu h}{1 + \nu(W + h)} \tag{2.50}$$

which corresponds to (2.24) for the M(n)/M/s(0) model.

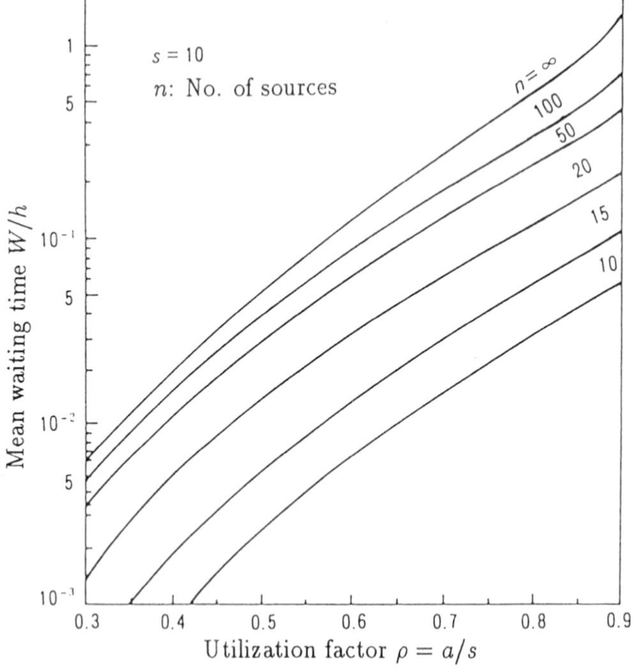

Fig.2.15 Mean waiting time of M(n)/M/s

[Example 2.3] Suppose that n terminals are connected to a TSS (time sharing service) computer, each originates transactions at random with mean interval ν^{-1} sec in the idle condition, and the transactions have exponentially distributed lengths [4, p.115].

The number of transactions handled per unit time is called the *throughput*, which is a performance measure of computer systems. The throughput is equivalent to the arrival rate λ, which is calculated from (2.50) by

$$\lambda = \frac{n\nu}{1+\nu(W+h)}. \tag{2.51}$$

If $n = 30$, $\nu^{-1} = 20\,\text{sec}$, $h = 1\,\text{sec}$, and a single CPU (central processing unit) is used, then we have $W = 9.2559\,\text{sec}$, from (2.49) with $s = 1$. Hence, the throughput is $\lambda = 0.9915\,\text{trans/sec}$ from (2.51), and the offered load is $a = 0.9915\,\text{erl}$ from (2.50). Therefore, the CPU utilization factor is $\rho = a/s = 99.15\,\%$.

If two CPUs are used, we have $W = 0.6758\,\text{sec}$ from (2.49) with $s = 2$. Thus, the waiting time is reduced and service is improved, but the throughput is $\lambda = 1.3840\,\text{trans/sec}$, and the CPU utilization factor is decreased to $\rho = 1.3840/2 = 69.20\,\%$.

2.3.5 M/M/$s(m)$

Letting $n \to \infty$ and $\xi \to 0$ in (2.49), we get

$$\begin{aligned}
W &= M(0)\frac{h}{s}\left(\frac{1}{1-\rho}+\frac{m\rho^m}{1-\rho^m}\right)\\
B &= P_0\frac{a^s}{s!}\rho^m\\
M(0) &= P_0\frac{a^s}{s!}\frac{1-\rho^m}{1-\rho}\\
M(t) &= M(0)e^{-st/h}\sum_{r=0}^{m-1}\frac{1-\rho^{m-r}}{1-\rho^m}\frac{(st/h)^r}{r!}
\end{aligned} \tag{2.52}$$

where $\rho = a/s$ and

$$P_0 = \left(\sum_{r=0}^{s-1}\frac{a^r}{r!}+\frac{a^s}{s!}\frac{1-\rho^{m+1}}{1-\rho}\right)^{-1}.$$

A numerical example is shown in Figure 2.16. It should be noted that the system with infinite waiting room diverges, and the waiting time becomes infinite as $\rho \to 1$. On the other hand, with finite waiting room, a steady state exists even for $\rho \geq 1$.

[Example 2.4] Consider a packet transmission system. If the packets with geometrically distributed length with mean l bits, are transmitted at speed v b/s, the transmission time (service time) can be approximated to an exponential distribution with mean $h = (l/v)$ sec. Assuming that the packets arrive at random, and a buffer of m packets is provided, the M/M/$s(m)$ model is applicable.

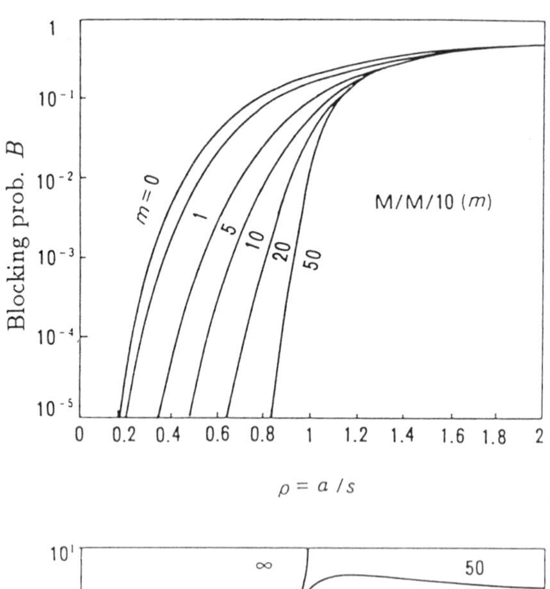

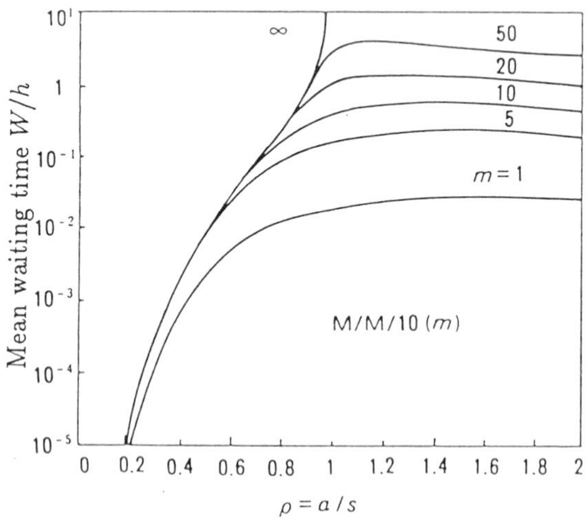

Fig.2.16 Numerical example of M/M/$s(m)$

For example, letting $l = 1200$ bits and $v = 2400$ b/s, we have $h = 1200/2400 = 0.5$ sec. If the arrival rate is $\lambda = 6$ packets/sec and the number of lines is $s = 4$, we have the traffic load $a = \lambda h = 6 \times 0.5 = 3$ erl. An example calculated by (2.52) is shown in Table 2.1.

The voice packets are delay sensitive but somewhat loss tolerable, so that a discarding (blocking) probability, say $B = 10^{-3}$, is allowed. Therefore, the buffer size $m = 10$ may be used. On the other hand, however, the data packets are delay tolerable but loss sensitive, so that say $B \leq 10^{-6}$ is required, and therefore $m \geq 40$ should be provided. If a sufficiently large buffer is provided, the *flow control* scheme is to be used which controls the input of data for preventing the system from overloading with $\rho \geq 1$ [7, p.94].

Table 2.1 Calculated example for M/M/$s(m)$

$a = 3\,\text{erl},\quad s = 4,\quad h = 0.5\,\text{sec}$

m	W [sec]	B	M (0.5 sec)
10	0.2090	0.007330	0.1622
20	0.2502	0.000404	0.1860
30	0.2543	0.000023	0.1873
40	0.2547	0.000001	0.1874

Exercises

[1] Suppose that in a local telephone office serving 10000 subscribers, the average subscriber calling rate (originating traffic load) is 0.04 erl, for which 10 % is directed to a toll office.

(1) Design the trunk circuits to the toll office with a blocking probability no greater than 1%.

(2) Estimate the blocking probability when the traffic is doubled under overload condition.

[2] A key telephone system with 4 telephone sets is connected to a central office via 2 subscriber lines. Suppose that each telephone set originates 2 calls per hour at random, and is used for 3 minutes (exponentially) on average. Then, calculate

(1) The blocking probability and the efficiency of the subscriber lines, when the system operates on a non-delay (loss) basis.

(2) The mean waiting time and the subscriber line utilization factor, when the system operates on a delay basis with infinite waiting room.

[3] Consider a computer center with 20 TSS terminals and 600 users. Assuming that each user uses the terminal once a day (8 hours) at random for 30 minutes (exponentially) on average, obtain

(1) The waiting probability and the mean waiting time for the terminal.

(2) The probabilities for the waiting time to exceed 6 minutes in FIFO and RSO, respectively.

[4] Suppose that packets with geometrically distributed length of mean 1200 bits, arrive at random at the rate of 6 packets/sec, and are transmitted over two 4800 b/s lines.

(1) Design the buffer size for keeping the discarding rate (blocking probability) no greater than 10^{-5}.

(2) With the buffer designed in (1), calculate the mean waiting time and the probability of the waiting time exceeding 1 second.

Chapter 3

NON-MARKOVIAN MODELS

A teletraffic system with the interarrival time or service time not exponentially distributed, are called a non-Markovian model. This chapter presents the main results for non-Markovian models for which relatively simple solutions are obtained. We shall begin by introducing the renewal process.

3.1 Renewal Process

3.1.1 Residual Time Distribution

If the time interval X between consecutive occurrences of a certain event, say call arrival, is independently and identically distributed (*iid*) with a distribution function $F(x)$, the process $\{X\}$ is called a *renewal process.* The Poisson process is a special case of the renewal process in which $F(x)$ is the exponential distribution.

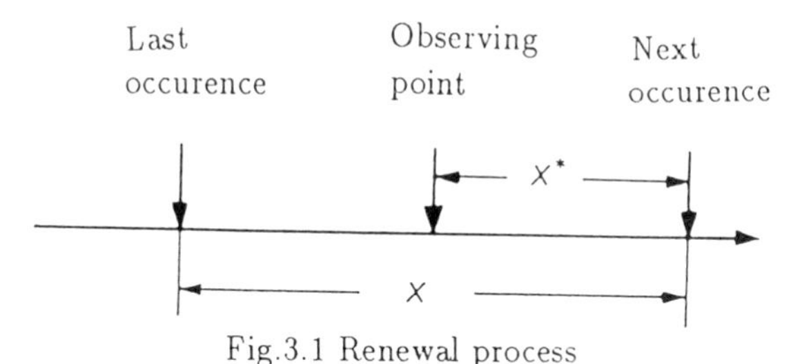

Fig.3.1 Renewal process

As shown in Figure 3.1, for a renewal process, the time interval X^* from an arbitrary observation point to the next occurrence instant is called *residual* (*life*) *time* or *forward recurrence time.* In this connection, the original occurrence interval X is referred to as the *lifetime.* The distribution function of X^* is given by

$$R(t) = \frac{1}{m}\int_0^t [1 - F(\xi)]d\xi \tag{3.1}$$

where $m = E\{X\}$ is the mean of X.

Equation (3.1) is derived as follows: Considering a sufficiently long time period $(0, T]$, the expected number of renewal intervals is T/m. Letting $f(x)$ be the *density function* of X, the probability that $x < X \leq x + dx$ is equal to $f(x)dx$, and hence the mean number of such intervals is $Tf(x)dx/m$. Thus, the expected total time occupied by such intervals in $(0, T]$ is given by

$$T_x = \frac{xTf(x)dx}{m}.$$

Let X' be the length of an observed interval. Then, since the observation point falls at random in time interval $(0, T]$, the probability that $x < X' \leq x + dx$, is the time fraction occupied by such intervals in $(0, T]$. Therefore, letting $f'(x)$ be the density function of X', we have

$$f'(x)dx = P\{x < X' \leq x + dx\} = \frac{T_x}{T} = \frac{x}{m}f(x)dx. \tag{3.2}$$

Since the observation point falls randomly over the observed interval X', given $X' = x$, the conditional probability that $t < X^* \leq t + dt$, is given by

$$P\{t < X^* \leq t + dt | X' = x\} = \begin{cases} \dfrac{dt}{x}, & x > t \\ 0, & x \leq t. \end{cases}$$

Hence, letting $r(t)$ be the density function of X^*, it follows that

$$r(t)dt = \int_{x=t}^{\infty} \left(\frac{dt}{x}\right) f'(x)dx = \frac{1}{m}[1 - F(t)]dt. \tag{3.3}$$

Integrating (3.3) yields (3.1).

The time from the last occurrence epoch to the observation point is called the *age* or *backward recurrence time*, and it is known that the age is also distributed according to (3.1).

[**Example 3.1**] If the lifetime X is distributed with the 2nd order hyper-exponential distribution, H_2, we have (See Table C.2 in Appendix C.)

$$\begin{aligned} F(x) &= 1 - [ke^{-\lambda_1 x} + (1-k)e^{-\lambda_2 x}] \\ m &= \frac{k}{\lambda_1} + \frac{1-k}{\lambda_2}. \end{aligned} \tag{3.4}$$

Hence, the residual time distribution is given from (3.1) by

$$R(t) = 1 - [k'e^{-\lambda_1 t} + (1-k')e^{-\lambda_2 t}] \tag{3.5}$$

where $k' = k/(m\lambda_1)$. Therefore, X^* is again distributed with H_2, but the parameter k' is different from k.

In particular, if $\lambda_1 = \lambda_2 = \lambda$, $F(x)$ in (3.4) becomes exponential, and the residual time distribution coincides with the lifetime distribution itself, so that

$$R(t) = F(t) = 1 - e^{-\lambda t}. \tag{3.6}$$

This results from the Markov property of the exponential distribution as described in Subsection 1.3.1.

3.1.2 Mean Residual Time

In general, let $F(x)$ be the distribution function of a continuous random variable $X \geq 0$. Then, we define the *Laplace-Stieltjes transform* (LST) of $F(x)$ by

$$f^*(\theta) = \int_0^\infty e^{-\theta x} dF(x) \equiv \int_0^\infty e^{-\theta x} f(x) dx \tag{3.7}$$

where

$$f(x) = \frac{dF(x)}{dx}$$

is the density function of X. The LST of the distribution function $F(x)$ is equivalent to the *Laplace transform* (LT) of the density function $f(x)$, and hence the notation $f^*(\theta)$ is used for the LST of $F(x)$ in this book.

The LST of the residual time distribution $R(t)$ is given by

$$r^*(\theta) = \frac{1 - f^*(\theta)}{m\theta} \tag{3.8}$$

where $f^*(\theta)$ is the LST of the lifetime distribution $F(x)$. (Use the integration formula for LST in Appendix C [T5].) From (3.8) we have the mean residual time,

$$R = -\lim_{\theta \to 0} \frac{dr^*(\theta)}{d\theta} = \frac{E\{X^2\}}{2m} \tag{3.9}$$

where $E\{X^2\} = \sigma^2 + m^2$ is the second moment of X, and σ^2 is the variance of X. (Use the moment formula and L'hospital theorem in Appendix C [T5].)

Define the *squared coefficient of variation* (SCV) by

$$C^2 = \frac{\sigma^2}{m^2}. \tag{3.10}$$

Then, we can express the mean residual time as

$$R = \frac{m}{2}(1 + C^2). \tag{3.11}$$

[**Example 3.2**] If the lifetime is distributed with the phase k Erlangian distribution, E_k, we have

$$F(x) = 1 - e^{-\lambda x} \sum_{j=0}^{k-1} \frac{(\lambda x)^j}{j!} \tag{3.12}$$

where $m = 1/\lambda$ and $C^2 = 1/k$. The LST of $F(x)$ is given by (See Table C.2 in Appendix C.)

$$f^*(\theta) = \left(\frac{k\lambda}{\theta + k\lambda} \right)^k. \tag{3.13}$$

From (3.8), we have the LST of the residual time distribution,

$$r^*(\theta) = \frac{\lambda}{\theta} \left[1 - \left(\frac{k\lambda}{\theta + k\lambda} \right)^k \right] \tag{3.14}$$

and from (3.11) the mean residual time,

$$R = \frac{1 + C^2}{2\lambda}. \tag{3.15}$$

In particular, if $k = 1$, the lifetime and the residual time are both identically and exponentially distributed with LST,

$$f^*(\theta) = r^*(\theta) = \frac{\lambda}{\theta + \lambda}. \tag{3.16}$$

In this case, we have $C^2 = 1$ and $m = R = 1/\lambda$, *i.e.* the mean lifetime and the mean residual time are identical. Although this seems a contradiction, it will be understood by noting that an observation point has a higher probability of falling in a longer lifetime than of falling in a shorter lifetime.

3.1.3 Rate Conservation Law

Consider a system in which calls arrive in a renewal process at rate λ and require the exponential service time with mean μ^{-1}. Letting P_j be the state probability that j calls exist in the steady state, and Π_j be that just prior to the call arrival, we have the *rate conservation law*,

$$\begin{aligned} &\text{Loss system}: \quad \lambda \Pi_{j-1} = j\mu P_j, \ j \le s \\ &\text{Delay system}: \quad \begin{cases} \lambda \Pi_{j-1} = j\mu P_j, \ j \le s \\ \lambda \Pi_{j-1} = s\mu P_j, \ j > s. \end{cases} \end{aligned} \tag{3.17}$$

For the loss system, the relation is interpreted as follows: Denoting the state of j calls existing in the system by S_j, since Π_{j-1} is the probability of S_{j-1} just prior to call arrival and λ is the call arrival rate, the left hand side of the equation represents the rate of $S_{j-1} \to S_j$. On the other hand, since $j\mu$ is the rate that a call terminates and P_j is the probability of j calls existing, the right hand side represents the rate of $S_j \to S_{j-1}$. Thus, both sides are balanced in the steady state, by the rate-up=rate-down.

The same interpretation applies to the delay system except that the termination rate for a call becomes $s\mu$ for $j > s$, since only s calls in service can be terminated.

3.2 Poisson Input General Service Time Models

It is known that multi-server loss systems with Poisson input and general service time, M/G/s(0), are equivalent to M/M/s(0), and the blocking probability is given by the Erlang B formula. Furthermore, the blocking probability for M(n)/G/s(0) is also given by the Engset loss formula [2]. These properties are referred to as the *robustness of service time.*

However, no exact solutions have yet been obtained for a general delay system, M/G/s, with Poisson inputs and general service time. As special cases, we shall present the solutions for M/G/1 in this section, and for M/D/s in the next section.

3.2.1 M/G/1

Consider the M/G/1 system with infinite buffer Poisson input at rate λ, and single server having a general service time distribution. It is known that the system has a steady state, if and only if the offered traffic load $a = \rho = \lambda h < 1$ erl, where h is the mean service time and ρ the utilization factor. This can be understood intuitively because a single server can serve 1 erl, at maximum.

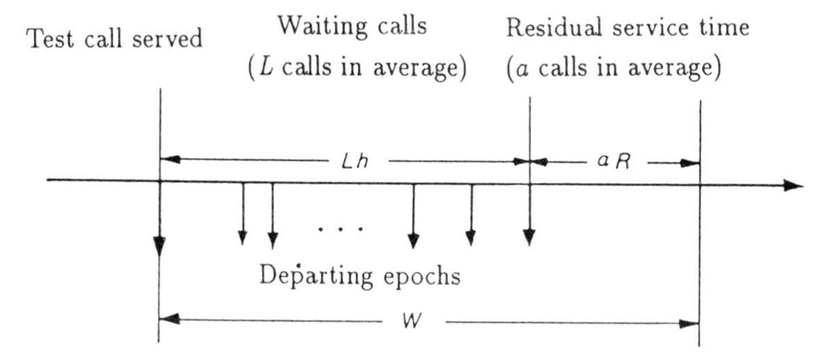

Fig.3.2 Mean waiting time for M/G/1

Choose an arbitrary call to be the *test call* as shown in Figure 3.2. The probability that the server is busy when the test call arrives, is a from Property (3) in Subsection 1.2.1 and the PASTA. The time until the call in service is terminated, is the residual service time. Hence, letting R be the mean residual time, L the mean number of waiting calls, and W the mean waiting time, with FIFO discipline, we have the relation,

$$W = aR + Lh. \tag{3.18}$$

In the right hand side, the first term corresponds to the mean time for a call in service, if any, to be terminated, and the second term corresponds to that for calls waiting ahead of the test call, to have been served. Using Little formula, $L = \lambda W$, and solving (3.18), we obtain

$$W = \frac{R}{1-\rho}. \tag{3.18a}$$

From (3.11), we have the mean residual time,

$$R = \frac{h}{2}(1 + C_s{}^2) \tag{3.19}$$

where $C_s{}^2 = \sigma_s^2/h^2$ is the SCV of the service time, with its variance σ_s^2. Using (3.19) in (3.18a), we obtain the *Pollaczek-Khintchine formula*,

$$W = \frac{\rho}{1-\rho}\frac{1+C_s{}^2}{2}h. \tag{3.20}$$

3.2.2 Embedded Markov Chain

In stochastic processes, a time instant at which the Markov property holds, is called the *renewal point.* (See Subsection 1.3.1.) For the M/G/1 system, the *departure epoch* at which a call is terminated and leaves the system, becomes the renewal point. The reason is as follows:

Let k be the number of calls present in the system just after a call departure epoch in the steady state. Then,

(1) If $k = 0$, since the system is empty when the next call arrives, the call is immediately served, and the number of calls present just after its departure is equal to the number of arrivals during its service time.

(2) If $k \geq 1$, since there are k calls waiting in the queue, the call at the top of the queue enters service, and the number of calls present at its departure is equal to k plus the number of arrivals during its service time.

In either case, because of the memory-less property of Poisson arrivals , the number of calls present just after the next (*future*) departure epoch is dependent only on the number k of calls existing just after the last (*present*) departure epoch and independent of the previous (*past*) progress, thus the Markov property holds.

A Markov process with discrete state space is called a *Markov chain*. In a Markov chain, all time epochs at which the state changes, are renewal points. On the other hand, a stochastic process is called an *embedded Markov chain*, if the renewal points are embedded or hidden in particular time epochs such as call departure in M/G/1 systems.

In what follows, M/G/1 systems are analyzed using the theory of embedded Markov chain.

3.2.3 State Probability at Call Departure

Let Π_j^* be the probability that j calls exist just after a call departure in the steady state. Then, we have the relation,

$$\Pi_j^* = p_j \Pi_0^* + \sum_{k=1}^{j+1} p_{j-k+1} \Pi_k^*, \; j = 0, 1, \cdots \tag{3.21}$$

where p_j is the probability that j calls arrive during the service time.

Equation (3.21) is derived as follows: The first term in the right hand side is the probability of j calls arriving when $k = 0$ corresponding Item (1) above. The second term is the probability of $(j - k + 1)$ calls arriving when $1 \leq k < j + 1$ corresponding to Item (2) (one call is the departing call itself). Thus, the right hand side probability is equal to Π_j^* in the left hand side. (See Figure 3.3.)

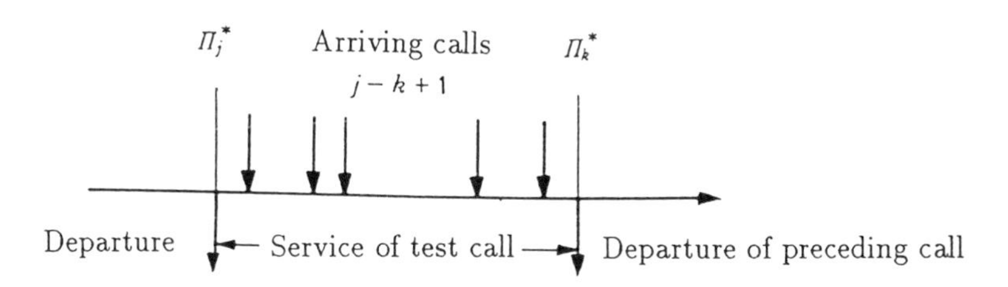

Fig.3.3 Embedded Markov chain

Since the conditional probability that j calls arrive in the service time, given the service time is equal to x, is

$$p_j(x) = \frac{(\lambda x)^j}{j!} e^{-\lambda x}$$

from the *total probability theorem* (See Appendix C [T2]), we have

$$p_j = \int_0^\infty \frac{(\lambda x)^j}{j!} e^{-\lambda x} dB(x) \tag{3.22}$$

where $B(t)$ is the service time distribution.

Define the *probability generating function* of Π_j^* by

$$g(z) = \sum_{j=0}^\infty z^j \Pi_j^*.$$

Multiplying (3.21) by z^j and summing, we have

$$g(z) = \Pi_0^* \sum_{j=0}^\infty z^j p_j + \sum_{j=1}^\infty \Pi_j^* \sum_{k=0}^\infty z^{j+k-1} p_k. \tag{3.23}$$

Noting $\lim_{z\to 1} g(z) = 1$, and

$$\sum_{j=0}^\infty z^j p_j = \int_0^\infty \sum_{j=0}^\infty z^j \frac{(\lambda x)^j}{j!} e^{-\lambda x} dB(x) = b^*(\lambda - z\lambda) \tag{3.24}$$

from (3.23) we have

$$g(z) = (1-a)\frac{(1-z)b^*(\lambda - z\lambda)}{b^*(\lambda - z\lambda) - z} \tag{3.25}$$

where $b^*(\theta)$ is the LST of the service time distribution function $B(t)$. We can calculate Π_j^*, by using the inversion formula (See Appendix C [T6].),

$$\Pi_j^* = \frac{1}{j!} \lim_{z\to 0} \frac{d^j}{dz^j} g(z) \equiv \frac{1}{j!} g^{(j)}(0). \tag{3.26}$$

Denoting the state probability at call arrival in the steady state by Π_j, it is known that $\Pi_j = \Pi_j^*$ in a stochastic process with discrete state space whose state transitions occur only by one step at a time [4, pp.185-188], and this is the case for the M/G/1 model under consideration. Moreover, letting P_j be the state probability seen by an outside observer, from the PASTA we have the relation,

$$P_j = \Pi_j = \Pi_j^*. \tag{3.27}$$

3.2.4 Waiting Time Distribution

Let $W(t)$ be the waiting time distribution of the test call, $w^*(\theta)$ its LST, and $f^*(\theta)$ the LST of its system time T (waiting time + service time). Then, since the waiting time and the service time are mutually independent, we have (by the *convolution formula* in Appendix C [T5])

$$f^*(\theta) = w^*(\theta)b^*(\theta). \tag{3.28}$$

Let N be the number of calls present just after the test call departure. Then, N is the same as the number of Poisson arrivals in T, because of the FIFO discipline. Similarly to (3.24), we have the probability generating function of N,

$$\hat{g}(z) = f^*(\lambda - z\lambda) \tag{3.28a}$$

which should be the same as $g(z)$ in (3.25). Therefore, setting $\hat{g}(z) = g(z)$ and $\theta = (1-z)\lambda$, we obtain the LST of the waiting time distribution function,

$$w^*(\theta) = \frac{(1-a)\theta}{\theta - \lambda[1 - b^*(\theta)]}. \tag{3.29}$$

Denote the residual time distribution function of the service time by $R(t)$, and its LST by $r^*(\theta)$. Then, noting (3.8) and rearranging (3.29), we obtain

$$w^*(\theta) = (1-a)\sum_{j=0}^{\infty} a^j [r^*(\theta)]^j. \tag{3.30}$$

Letting $R^{*j}(t)$ be the j hold convolution of $R(t)$, noting that its LST is $[r^*(\theta)]^j$, and inverting (3.30), we obtain the *Beneš formula*,

$$W(t) = (1-a)\sum_{j=0}^{\infty} a^j R^{*j}(t). \tag{3.31}$$

It can be shown that $W(t)$ satisfies the *Volterra integral equation*,

$$W(t) = P_0 + \lambda \int_0^t W(t-x)[1 - B(x)]dx \tag{3.31a}$$

where $P_0 = 1 - a$ is the probability that the system is empty at an arbitrary time in the steady state.

[Example 3.3] Consider an M/D/1 system with arrival rate λ, and fixed service time h with SCV $C_s^2 = 0$. Then, the mean waiting time is given from (3.20) by

$$W = \frac{1}{2}\frac{a}{1-a}h \tag{3.32}$$

where $a = \lambda h$ is the offered load. It should be noted that (3.32) is a half of the value for M/M/1 by (2.38),

$$W = \frac{a}{1-a}h. \tag{3.32a}$$

The residual time of the fixed service time has a *uniform distribution* function,

$$R(t) = \begin{cases} 1, & t \geq h \\ \frac{t}{h}, & 0 < t < h \\ 0, & t \leq 0, \end{cases} \tag{3.33}$$

and its j hold convolution is given by [5, p.14]

$$R^{*j}(t) = \frac{1}{j!}\left(\frac{t}{h}\right)^j, \quad 0 < t \leq h. \tag{3.33a}$$

Using (3.33a) in (3.31) yields the waiting time distribution function for M/D/1,

$$W(t) = (1-a)\sum_{j=0}^{\infty}\frac{a^j}{j!}\left(\frac{t}{h}\right)^j = (1-a)e^{\lambda t}, \quad 0 < t \leq h. \tag{3.34}$$

3.2.5 M/G/1(m)

Next, let us consider the case with finite waiting room with m positions. Using the same notation as before, and noting that $\Pi_j^* = 0$ for $j \geq m+1$ since a departing call can leave behind at most m calls, we have from (3.21) the recurrence formula,

$$\Pi_{j+1}^* = \left(\Pi_j^* - p_j\Pi_0^* - \sum_{k=0}^{j} p_{j-k+1}\Pi_k^*\right) p_0^{-1}, \quad j = 0, 1, \cdots, m-1 \tag{3.35}$$

where p_j is defined by (3.22). Noting that p_j is the coefficient of z^j in (3.24), it is calculated by

$$p_j = \frac{1}{j!}\lim_{z\to 0}\frac{\partial^j}{\partial z^j}b^*([1-z]\lambda) \tag{3.36}$$

for which some examples are given below:

(1) For constant service time (D), since the LST is given by (See Table C.2 in Appendix C.)

$$b^*(\theta) = e^{-h\theta} \tag{3.37}$$

from (3.36) we have

$$p_j = e^{-a}\frac{a^j}{j!}, \quad j = 0, 1, \cdots. \tag{3.38}$$

(2) When the service time is distributed with E_k (Erlangian distribution), with parameters $k = 1/C_s^2$ and $\mu = h^{-1}$, we have the LST of the form of (3.13) and

$$p_0 = \left(\frac{k}{k+a}\right)^k, \quad p_j = \frac{(k+j-1)a}{j(k+a)}p_{j-1}, \quad j = 1, 2, \cdots \tag{3.39}$$

where $a = \lambda h$ is the offered load.

(3) If the service time is distributed with H_2 (hyper-exponential distribution), the LST is given by (See Table C.2 in Appendix C.)

$$b^*(\theta) = \frac{k}{\theta + \mu_1} + \frac{1-k}{\theta + \mu_2} \tag{3.40}$$

where k, μ_1 and μ_2 are determined by the first 3 moments of the service time. (See (3.101).) Using (3.40) in (3.36), we have

$$\begin{aligned} p_j &= kp_{1j} + (1-k)p_{2j}, \quad j = 0, 1, \cdots \\ p_{ij} &= \frac{k\lambda}{\lambda + \mu_i}p_{i,j-1}, \quad p_{i0} = \frac{\mu_j}{\lambda + \mu_i}, \quad i = 1, 2. \end{aligned} \tag{3.40a}$$

Now, we shall present an algorithm for calculating the state probability from which the performance measures such as blocking probability B, the waiting probability $M(0)$ and the mean waiting time W, are calculated.

Corresponding to (3.27), we have

$$\begin{aligned} P_j &= \Pi_j = \Pi_j^*, \quad j = 0, 1, \cdots, m \\ P_{m+1} &= \Pi_{m+1}. \end{aligned} \tag{3.41}$$

In (3.35), setting $C_j = \Pi_j^*/\Pi_0^*$, we have

$$C_{j+1} = \left(C_j - p_j + \sum_{k=1}^{j} p_{j-k+1}C_k\right)p_0^{-1}, \quad j = 0, 1, \cdots, m-1. \tag{3.42}$$

By recursively solving (3.42) for C_j, set

$$C = 1 + \sum_{j=1}^{m} C_j. \tag{3.43}$$

Noting (3.41), we have

$$P_j = \frac{C_j}{1 + aC}, \quad j = 0, 1, \cdots, m \tag{3.44}$$

using which, we obtain

$$B = P_{m+1} = 1 - \frac{C}{1 + aC} \tag{3.45}$$

$$M(0) = 1 - P_0 - B. \tag{3.46}$$

The mean number of waiting calls is given by

$$L = \sum_{j=1}^{m+1} (j - 1) P_j \tag{3.47}$$

and from the Little formula, we obtain

$$W = P_0 \left[mC - a^{-1} \sum_{k=1}^{m} (m - k + 1) C_k \right] h. \tag{3.48}$$

[Example 3.4] In M/D/1(m) system with $a = 0.5\,\text{erl}$, $h = 1\,\text{sec}$ and $m = 2$, from (3.44) we have

$$P_0 = 0.5136, \quad P_1 = 0.3332, \quad P_2 = 0.1259, \quad P_3 = B = 0.0272.$$

Hence, from (3.46) and (3.48), respectively, we obtain

$$M(0) = 0.4591, \quad W = 0.3608\,\text{sec}.$$

Figure 3.4 shows an example of calculation for various a and m.

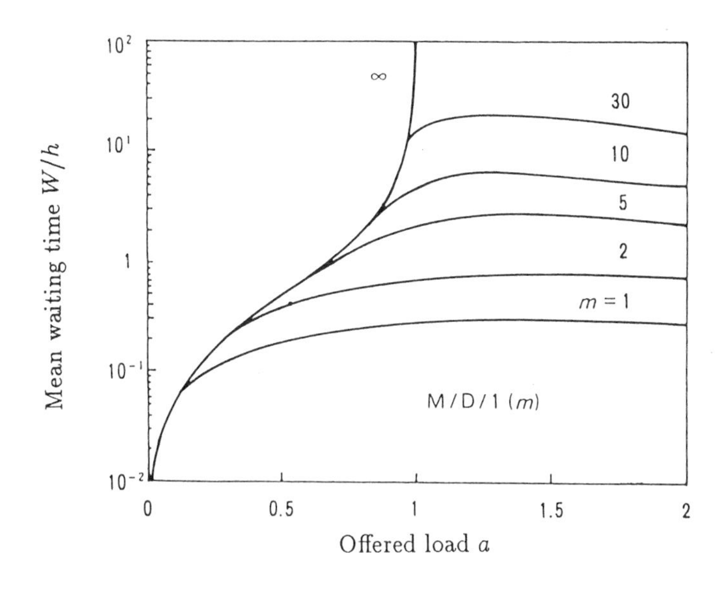

Fig.3.4 Mean waiting time of M/D/1(m)

3.3 Poisson Input Constant Service Time Model

3.3.1 M/D/s

Consider the M/D/s system with constant service time h, and let $N(t)$ be the number of calls present in the system at time t. Then, all calls in service at time t should have left the system at time $(t + h)$ because of the constant service time. Hence, the event $\{N(t + h) = j\}$, $j = 1, 2, \cdots$, results from one of the following cases:

(1) $N(t) \leq s$, and j calls arrive in time interval $(t, t + h]$. In this case, no calls are waiting at time t, and only calls arriving in $(t, t + h]$ exist at time $(t + h)$.

(2) $N(t) = s + k$, $k \geq 1$, and $(j - k)$ calls arrive in $(t, t + h]$. In this case, k calls waiting at time t still exist at time $(t + h)$.

Assume the existence of a steady state as $t \to \infty$, and let P_j be the steady state probability that j calls exist in the system. Then, we have

$$\begin{aligned} P_0 &= p_0 \sum_{k=0}^{s} P_k \\ P_j &= p_j \sum_{k=0}^{s} P_k + \sum_{k=1}^{j} p_{j-k} P_{s+k}, \quad j = 1, 2, \cdots \end{aligned} \tag{3.49}$$

where p_j is the probability that j calls arrive in the constant service time h, and is given by

$$p_j = \frac{a^j}{j!} e^{-a} \tag{3.50}$$

with the offered load $a = \lambda h$ and the arrival rate λ.

Equation (3.49) is interpreted as follows: The right hand side of the first equation represents the probability that $N(t) \le s$ and no calls ($j = 0$) arrive in $(t, t+h]$ to result in $N(t+h) = 0$ in Case (1) above . The right hand side of the second equation is the probability that either j (≥ 1) calls arrive in Case (1) or $(j-k)$ calls arrive in Case (2) in $(t, t+h]$ to result in $N(t+h) = j$. In the steady state, the both sides of the equations are balanced, respectively. In fact, it is known that the steady state exists if and only if $a < s$.

3.3.2 State Probability Generating Function

Define the probability generating function of P_j by

$$g(z) = \sum_{j=0}^{\infty} z^j P_j$$

and set

$$g_r(z) = \sum_{j=0}^{r} z^j P_j, \quad Q_r = \sum_{j=0}^{r} P_j. \tag{3.51}$$

Then, from (3.49) we can derive

$$g(z) = \frac{g_s(z) - z^s Q_s}{1 - z^s e^{a(1-z)}}. \tag{3.52}$$

The denominator of the right hand side of (3.52) has s zeros at z_r, $r = 0, 1, \cdots, s-1$, which are the roots of the transcendental equation,

$$1 - z^s e^{a(1-z)} = 0. \tag{3.53}$$

One root is clearly $z_0 = 1$, and the other $(s-1)$ roots may be obtained as follows:

By setting $z = \gamma e^{j\omega}$ with $j = \sqrt{-1}$ and $\rho = a/s$, and by separating the real and imaginary parts of (3.53), we have

$$\begin{aligned} \gamma &= \exp(\rho\gamma\cos\omega - \rho) \\ \omega &= \rho\gamma\sin\omega + \frac{2\pi r}{s}. \end{aligned} \tag{3.54}$$

Iterating (3.54) for $r = 1, 2, \cdots, s-1$ yields γ_r and ω_r for z_r.

Since $g(z)$ is non-singular (having no poles) for $|z| \le 1$, the numerator of the right hand side of (3.52) should have also zeros at z_r, $r = 1, 2, \cdots, s-1$. Hence, the numerator may be written in the form,

$$g_s(z) - z^s Q_s = -K \prod_{r=1}^{s-1} (z - z_r) \tag{3.55}$$

where K is the unknown constant. Substituting (3.55) into (3.52), and using the relation $g(z) \to 1$ as $z \to 1$, to determine K, using the *L'Hospital theorem* (See Appendix C [T5].), we have the generating function,

$$g(z) = \frac{(1-z)(s-a)}{1 - z^s e^{a(1-z)}} \prod_{r=1}^{s-1} \frac{z - z_r}{1 - z_r}. \tag{3.56}$$

3.3.3 Mean Waiting Time

The mean number N of calls present in the steady state is given by (See moment formula in Appendix C [T6].)

$$N = \lim_{z \to 1} \frac{dg(z)}{dz} = \sum_{r=1}^{s-1} \frac{1}{1 - z_r} + \frac{s - (s-a)^2}{2(s-a)}. \tag{3.57}$$

Since the mean number of calls in service is a from Property (4) in Subsection 1.2.1, the mean number of waiting calls is $L = N - a$, and from the Little formula, the mean waiting time is given by

$$W = \left[\sum_{r=1}^{s-1} \frac{1}{1 - z_r} + \frac{a^2 - s(s-1)}{2(s-a)} \right] \frac{h}{a}. \tag{3.58}$$

Of the $(s-1)$ roots z_r, $r = 1, 2, \cdots, s-1$ calculated by (3.54), if s is an even number, $z_{s/2}$ is real and the rest are of $(s/2 - 1)$ conjugate pairs; if s is an odd number, the roots are of $(s-1)/2$ conjugate pairs. In either case, the first term in

[ ] of (3.58) becomes real by virtue of the conjugate pairs. Figure 3.5 shows an example calculated by (3.58).

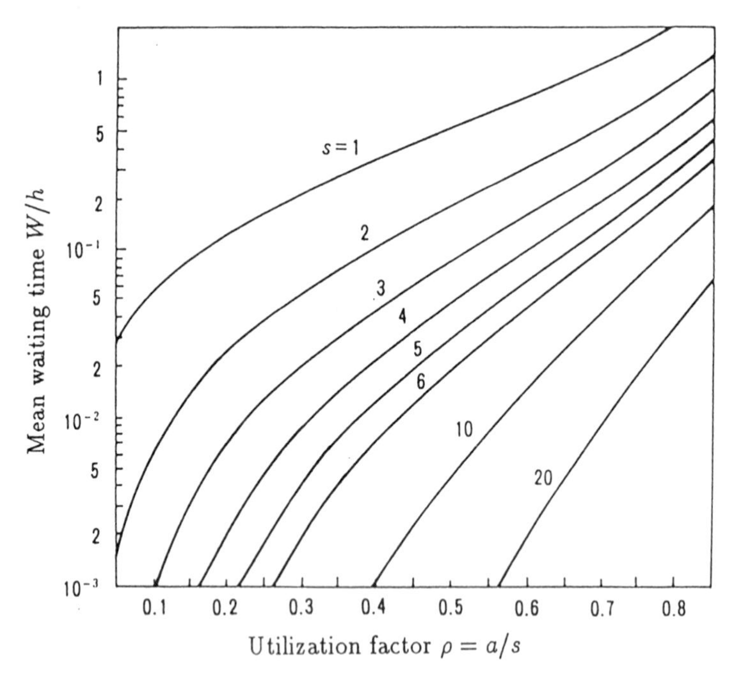

Fig.3.5 Mean waiting time of M/D/s

Applying the complex integral to modify (3.58), we have *Crommelin-Pollaczek formula* [13],

$$W = h\sum_{i=1}^{\infty} e^{-ia}\left[\sum_{k=is}^{\infty}\frac{(ia)^k}{k!} - \frac{s}{a}\sum_{k=is+1}^{\infty}\frac{(ia)^k}{k!}\right]. \tag{3.59}$$

It should be noted that although (3.59) does not involve the complex roots, the convergence is slow.

[Example 3.5] Consider the system shown in Figure 1.2 in Chapter 1, which transmits packets with constant length $l = 2400$ bits arriving at random at rate $\lambda = 3$ packets/sec. It is assumed that a sufficiently large buffer is provided.

In the case of 4 lines of speed $v = 2400$ b/s, the mean transmission (service) time of a packet is $h = l/v = 1$ sec, and the offered load is $a = \lambda h = 3$ erl. With $s = 4$, by iterating (3.54) we obtain

$$\begin{aligned}\gamma_1 &= 0.4287, & \gamma_2 &= 0.3605, & \gamma_3 &= 0.4287\\ \omega_1 &= -1.2643, & \omega_2 &= 0, & \omega_3 &= 1.2643\end{aligned}$$

where ω_r is measured in radian. Letting $z_k = \gamma_k \exp(j\omega_k)$, from (3.58) we have $W = 0.2670$ sec.

In general, the term *response time* is used as a measure of GOS, which is defined as the time from the data (packet) origination to receipt at the destination, *i.e.* time for waiting + transmission + propagation. If neglecting the propagation time, the response time is equivalent to the system time. In this example the mean response time is $T_R = 1.2670$ sec.

On the other hand, when using a single line with 4-fold increased speed, 9600 b/s, the transmission time is reduced to $h = 0.2$ sec. From (3.32), we have $W = 0.3750$ sec, and hence $T_R = 0.6250$ sec. Thus, with the single high-speed line, the mean response time is reduced to about a half of that for the above multi-line system.

3.3.4 Waiting Time Distribution

Let us find the waiting time distribution with the FIFO discipline in the steady state. Let $b(r)$ be the probability that r calls out of those present at an arbitrary instant, still exist after time x, $0 \le x < h$, has elapsed. Then, setting $t = kh + x$, where k is a non-negative integer, the waiting time distribution function is given by

$$W(kh+x) = b(ks+s-1), \quad k = 0, 1, \cdots; \; 0 \le x < h. \tag{3.60}$$

Equation (3.60) is derived as follows: Choose an arbitrary call, and let it be the test call. Suppose that the test call arrives at time t_0 in the steady state, and $(ks+s-1)$ calls exist at time (t_0+x). Since s calls in service depart the system during every time h, after time kh has been elapsed, the number of busy servers becomes $(s-1)$. Then, the test call enters service.

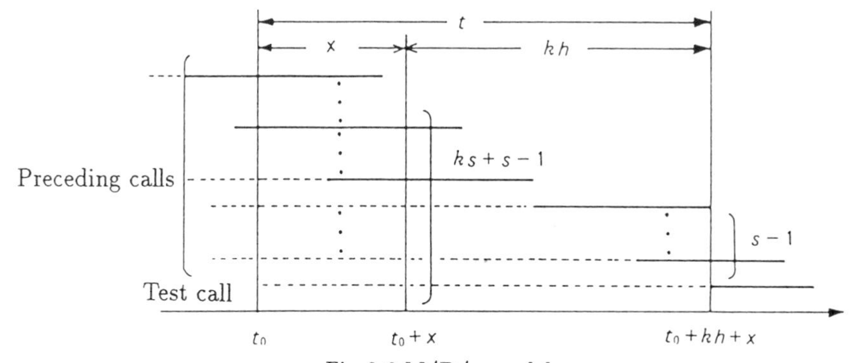

Fig.3.6 M/D/s model

Letting λ be the Poisson arrival rate, the probability that no greater than r calls exist at time (t_0+x) is given by

$$Q_r(t_0 + x) = e^{-\lambda x} \sum_{i=0}^{r} b(i) \frac{(\lambda x)^{r-i}}{(r-i)!}, \quad k = 0, 1, 2, \cdots \tag{3.61}$$

which should coincide with Q_r in (3.51) as $t_0 \to \infty$. Using this relation to obtain $b(r)$, we have from (3.60)

$$W(kh + x) = \sum_{j=0}^{k} \sum_{r=0}^{s-1} \frac{[-\lambda(jh + x)]^{(k-j+1)s-1-r}}{[(k-j+1)s - 1 - r]!} Q_r e^{\lambda(jh+x)}, \quad k = 0, 1, 2, \cdots. \tag{3.62}$$

In (3.51), Q_r is evaluated in terms of P_j which is calculated by

$$P_j = (-1)^{s-j+1} W(0^+) a_{s-j}, \quad 0 \le j \le s - 1 \tag{3.63}$$

where, $W(0^+)$ is the probability of no waiting, and is given by

$$W(0^+) = (s - a) \prod_{r=1}^{s-1} \frac{1}{1 - z_r} = Q_{s-1}. \tag{3.64}$$

The coefficient a_r is defined

$$\sum_{r=0}^{s} (-1)^r a_r z^{s-r} = \prod_{r=1}^{s-1} (z - z_k) \equiv y(z). \tag{3.65}$$

and is calculated by

$$a_r = \frac{(-1)^r}{k!} \lim_{z \to 0} \frac{d^r y(z)}{dz^r} \tag{3.66}$$

and we have

$$a_0 = 1, \quad a_1 = \sum_{i=0}^{s-1} z_i, \quad a_2 = \sum_{\substack{i=0 \\ (i<j)}}^{s-1} \sum_{j=0}^{s-1} z_i z_j, \quad \cdots, \quad a_s = z_0 z_1 \cdots z_{s-1}. \tag{3.67}$$

In particular, with $s = 1$, (3.62) reduces to

$$W(kh + x) = (1 - a) \sum_{j=0}^{k} \frac{[-\lambda(jh + x)]^{k-j}}{(k-j)!} e^{\lambda(jh+x)}, \quad k = 0, 1, 2, \cdots \tag{3.68}$$

which coincides with (3.34) for $k = 0$. Figure 3.7 shows a numerical example of the complementary distribution function $M(t) = 1 - W(t)$ calculated by (3.62).

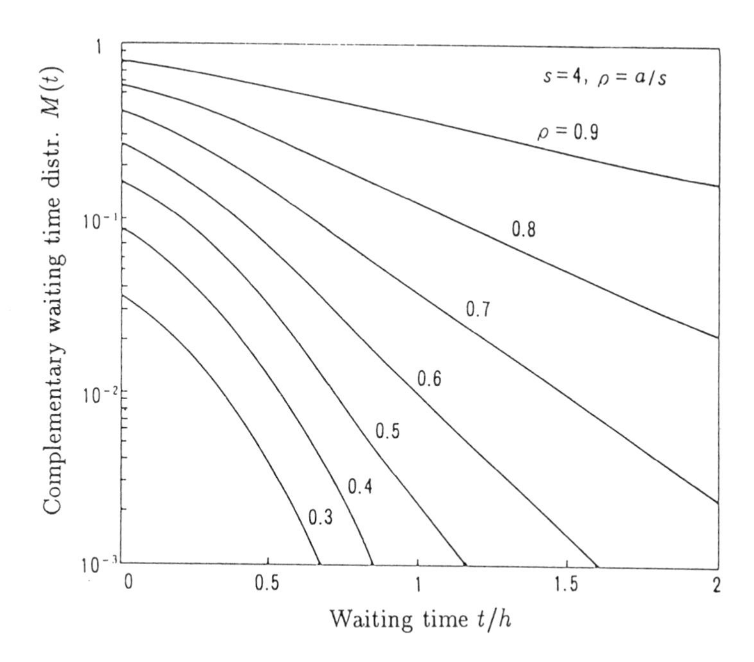

Fig.3.7 Waiting time distribution of M/D/s

[Example 3.6] In Example 3.5, let us calculate the probability that the response time is no greater than 3 sec.

For $s = 4$ and $v = 2400$ b/s, from (3.51) we have

$$Q_0 = 0.0338, \ Q_1 = 0.1412, \ Q_2 = 0.3155, \ Q_3 = 0.5096.$$

Since the packet transmission (service) time $h = 1$ sec is fixed, the probability that the waiting time is no greater than 2 sec (response time $-$ transmission time), is $W(2\,\text{sec}) = 0.0076$ from (3.62) with $s = 4$, $k = 2$ and $x = 0$.

In the case of using a single 9600 b/s line, since $h = 0.25$ sec and the allowable waiting time is 2.75 sec ($= 3 - 0.25$), we have the probability that the response time is no greater than 3 sec is $W(2.75\,\text{sec}) = 0.0020$ which is smaller than that for the multi-channel system. Thus, the single high-speed line is advantageous in service.

3.4 Renewal Input Exponential Server Models

If the input process is a renewal process, it is called *renewal input* and denoted by notation GI (general independent). This section deals with systems with renewal input and a multiple of servers having the same exponential service time.

3.4.1 GI/M/s(0)

Consider a multi-server loss system with renewal input and exponential service time. In this system, the call arrival epochs become renewal points, and constitute an embedded Markov chain.

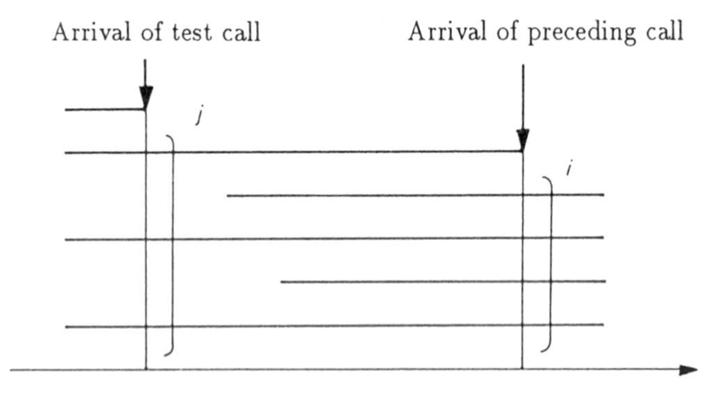

Fig.3.8 GI/M/s model

Mark the test call as before, and let Π_j be the probability that j calls exist just prior to the test call arrival in the steady state. Then, we have the relation,

$$\Pi_j = \sum_{i=j-1}^{s} p_{ij}\Pi_i, \quad j = 0, 1, \cdots, s \tag{3.69}$$

where p_{ij} is the transition probability that the state (number of calls existing in the system) changes from i to j between two consecutive arrival epochs, and given by

$$p_{ij} = \int_0^\infty \binom{i+1}{j} e^{-j\mu x}(1-e^{\mu x})^{i-j+1} dA(x), \quad 0 \leq j \leq i+1 \leq s \tag{3.70}$$

with $A(t)$ being the interarrival time distribution function, and μ^{-1} the mean service time. (See Figure 3.8.)

Equation (3.70) is interpreted as follows: If i calls exist just prior to the preceding call arrival epoch, the number of calls present just after the call arrival becomes $(i+1)$, and the right hand side of (3.70) represents the probability that j out of

$(i+1)$ calls survive just prior to the test call arrival epoch. (See the pure-death process in Appendix C.5.4.)

Letting $g(z)$ be the probability generating function of Π_j, from (3.69) and (3.70) we have

$$g(z) \equiv \sum_{j=0}^{s} z^j \Pi_j = \sum_{j=0}^{s-1} \Pi_j \int_0^\infty [1 - e^{-\mu x}(1-z)]^{j+1} dA(x)$$
$$+\Pi_s \int_0^\infty [1 - e^{-\mu x}(1-z)]^s dA(x). \tag{3.70a}$$

3.4.2 Blocking Probability

Here, we introduce the kth *binomial moment* of Π_j,

$$B_k \equiv \sum_{j=k}^{s} \binom{j}{k} \Pi_j = \frac{1}{k!} \lim_{z\to 1} \frac{\partial^k}{\partial z^k} g(z)$$
$$= \left[B_k + B_{k-1} - \binom{s}{k-1} \Pi_s \right] a^*(k\mu), \quad k = 1, 2, \cdots, s \tag{3.70b}$$

where $a^*(\theta)$ is the LST of $A(t)$. From (3.70b), we have the difference equation,

$$\frac{B_{k-1}}{C_{k-1}} - \frac{B_k}{C_k} = \binom{s}{k-1} \frac{\Pi_s}{C_{k-1}} \tag{3.70c}$$

where

$$\phi(\theta) = \frac{a^*(\theta)}{1 - a^*(\theta)}, \quad C_r = \prod_{i=1}^{r} \phi(i\mu), \quad C_0 = 1. \tag{3.71}$$

Summing side by side of (3.70c) from $k+1$ to s, yields

$$B_k = \Pi_s C_k \sum_{r=k}^{s} \binom{s}{r} \frac{1}{C_r}. \tag{3.72}$$

Using the relation $B_0 = 1$ in (3.72), we have the blocking probability (call congestion probability),

$$B = \Pi_s = \left[1 + \sum_{r=1}^{s} \binom{s}{r} \prod_{i=1}^{r} \frac{1}{\phi(i\mu)} \right]^{-1}. \tag{3.73}$$

From the definition in (3.70b), $B_{k-1} = \Pi_{s-1} + s\Pi_s$, and using this in (3.72), we obtain

$$\Pi_{s-1} = \frac{\Pi_s}{\phi(s\mu)}. \tag{3.73a}$$

Let λ be the arrival rate, and $a = \lambda/\mu$ the offered load. Applying the rate conservation law in (3.17), and (3.73a), the time congestion probability (probability of all servers busy) is given by

$$B_T = P_s = \frac{\lambda \Pi_{s-1}}{s\mu} = \frac{a\Pi_s}{s\phi(s\mu)}. \tag{3.74}$$

[Example 3.7] Suppose that trucks delivering components, arrive at a firm at fixed 10 min intervals. Assume that the time for unloading is exponentially distributed with mean 30 min, and trucks finding all parking places busy go away. Design the number s of parking places required for the component to be accepted with probability 95%.

This is modeled as D/M/s(0), and the LST of the fixed interarrival time (unit distribution) is given by (See Table C.2 in Appendix C.)

$$a^*(\theta) = e^{-\theta/\lambda}. \tag{3.75}$$

Since $\lambda = 1/(10\,\text{min})$ and $\mu = 1/(30\,\text{min})$, the offered traffic load is $a = \lambda/\mu = 3\,\text{erl}$, and the probability B that the all parking spots are occupied when a truck arrives is given by $B = 0.0334 < 0.05$ from (3.73), with $s = 5$ which is the required number ($B = 0.1086$ with $s = 4$). In this case, the probability that all the parking places are occupied as seen by an outside observer, is $B_T = 0.0862$ from (3.74), which is much greater than B above.

3.4.3 GI/M/s

Consider the corresponding delay system with infinite buffer and the FIFO discipline. Using similar notation as above, we have the steady state equation,

$$\Pi_j = \sum_{i=j-1}^{\infty} p_{ij}\Pi_i, \quad j = 0, 1, \cdots. \tag{3.76}$$

Note that since there is no limit for waiting calls, the summation extends to infinity. The transition probability p_{ij} is given by

$$p_{ij} = \int_0^\infty \frac{(s\mu x)^{i-j+1}}{(i-j+1)!} e^{s\mu x} dA(x), \quad s \le j \le i+1. \tag{3.77}$$

Given that i calls exist just prior to the preceding call arrival, $(i+1)$ calls exist at its arrival epoch, and (3.77) is the conditional probability that $(i-j+1)$ calls terminate during its service time, to result in j calls existing just prior to the test call arrival. (See Figure 3.8.)

Using (3.77) in (3.76), we have

$$\Pi_j = \Pi_s \omega^{j-s}, \quad j = s-1, s, \cdots \tag{3.78}$$

where Π_s is the boundary probability of exactly s calls existing in the system just prior to a call arrival, and is given by (See Subsection 3.4.5.)

$$\Pi_s = \left[\frac{1}{1-\omega} + \sum_{r=1}^{s} \binom{s}{r} \frac{s[1-a^*(r\mu)] - r}{[s(1-\omega)-r][1-a^*(r\mu)]} \prod_{i=1}^{r} \frac{1}{\phi(i\mu)} \right]^{-1}. \tag{3.79}$$

Here, $\phi(\theta)$ is defined in (3.71), and ω is called the *generalized occupancy* given as the root $(0 < \omega < 1)$ of the functional equation,

$$\omega = a^*([1-\omega]s\mu) \tag{3.80}$$

which may be calculated by iteration with the initial value $\omega^0 = \rho = \lambda/(s\mu)$.

The waiting probability is given by

$$M(0) = \sum_{j=s}^{\infty} \Pi_j = \frac{\Pi_s}{1-\omega}. \tag{3.81}$$

From the rate conservation law (3.17), the state probability seen by an outside observer is given by

$$P_j = \frac{a}{s} \Pi_{j-1}, \quad j = s, s+1, \cdots. \tag{3.82}$$

Using (3.82) we have the mean number of waiting calls,

$$L = \sum_{j=s}^{\infty} (j-s) P_j = \frac{a\Pi_s}{s(1-\omega)^2}. \tag{3.83}$$

Letting $h = \mu^{-1}$ be the mean service time, from the Little formula we have the mean waiting time,

$$W = M(0) \frac{h}{s(1-\omega)}. \tag{3.84}$$

3.4.4 Waiting Time Distribution

Let us derive the waiting time distribution with the FIFO discipline. Denoting by Q the number of waiting calls just prior to the test call arrival, the conditional probability that $Q = j$, given the call has to wait, is given by

$$\begin{aligned} P\{Q = j \mid Q > 0\} &= \frac{P\{(Q=j) \cap (Q>0)\}}{P\{Q>0\}} = \frac{\Pi_{s+j}}{M(0)} \\ &= (1-\omega)\omega^j, \quad j = 1, 2, \cdots. \end{aligned}$$

Since this is distributed geometrically as in (2.39) for M/M/1, replacing ρ in (2.41) by ω, we have the complementary distribution function,

$$M(t) = M(0) e^{-(1-\omega)st/h}. \tag{3.85}$$

[Example 3.8] Consider a test process where two types of components A and B arrive alternately each according to the Poisson process. If the test line for each type consists of s test members and the testing time is exponentially distributed, each test process is modeled as $E_2/M/s$.

If 4 components of each type arrive per hour, the arrival rate at each test line is $\lambda = 4\,\text{components/hr}$. With $s = 6$ members and mean test time $\mu^{-1} = 1\,\text{hr}$, we have from (3.80)

$$\omega = \left[\frac{2\lambda}{(1-\omega)s\mu + 2\lambda}\right]^2. \tag{3.86}$$

Solving this by iteration with the initial value $\omega^0 = \lambda/(s\mu) = 2/3$, we obtain $\omega = 0.5750$. Using this ω in (3.81) and (3.84) we have $M(0) = 0.2042$ and $W = 0.0801\,\text{hr} = 4.806\,\text{min}$. With FIFO discipline, the probability that the waiting time in a test line is greater than 6 min, is $M(6\,\text{min}) = 0.1583$ from (3.85).

If the types are not distinguished and tested in common, the system becomes an $M/M/s$ with $\lambda = 8$ components/hr and $s = 12$. Then, we have $W = 0.0350\,\text{hr} = 2.098\,\text{min}$ and $M(6\,\text{min}) = 0.0937$ from (2.38) and (2.41), thus the test process is considerably improved.

3.4.5 Derivation of Boundary Probability

Let us derive the boundary probability Π_s in (3.79) [4, p.269]. In addition to (3.77) for $s \le j \le i+1$, we have the transition probabilities p_{ij} for other ranges,

$$p_{ij} = \int_0^\infty \binom{i+1}{j} e^{-j\mu x}(1-e^{-\mu x})^{i+1-j}dA(x), \quad j \le i+1 < s \tag{3.77a}$$

$$p_{ij} = \int_0^\infty \int_0^x \frac{(s\mu y)^{i-s}}{(i-s)!} s\mu e^{-s\mu y} \binom{s}{j} e^{-j\mu(x-y)}[1-e^{-\mu(x-y)}]^{s-j} dy dA(x),$$
$$j < s \le i. \tag{3.77b}$$

Equation (3.77a) is the same as (3.70), while (3.77b) is derived as follows:

Let T_k be the time instant just prior to the kth call arrival in the steady state, and $N(t)$ the number of calls present in the system at time t. Then, taking the time origin at T_k, given $N(T_k) = i \ge s$, the event $\{N(T_{k+1}) = j\}$, $j < s$ results from events A and B below:

A: The kth call arrives at $T_k + 0$, and we have $N(T_k + 0) = i + 1$. Then, $(i - s)$ calls terminate in $(0, y]$ and a call terminates in $(y, y + dy]$ to result in $N(y + dy) = s$, where $T_k = 0 < y < T_{k+1}$. In this interval, since s calls are served with exponential service time with mean μ^{-1}, the number of terminating calls follows the Poisson distribution with rate $s\mu y$, and the probability of one call terminating in $(y, y+dy]$ is $s\mu y dy$. Hence, the probability of event A is given by

$$P\{A\} = \frac{(s\mu y)^{i-s}}{(i-s)!} e^{-s\mu y} s\mu y dy.$$

B: In the residual time interval $(y+dy, T_{k+1}]$, $(s-j)$ calls terminate to result in $N(T_{k+1}) = j$. Since the system state constitutes a pure-death process as in (3.70), the probability of event B is given by

$$P\{B\} = \binom{s}{j} e^{-j\mu(x-y)}[1-e^{-\mu(x-y)}]^{s-j} dA(x).$$

Since events A and B are independent, integrating $P\{A\}P\{B\}$ yields (3.77b).

Using (3.77), (3.77a) and (3.77b) in (3.69) for respective ranges, and introducing

$$U(z) = \sum_{j=0}^{s-1} z^j \Pi_j, \quad U_j = \frac{1}{j!} \lim_{z \to 1} \frac{\partial^j}{\partial z^j} U(z), \quad U_0 = U(1) = 1 - \frac{\Pi_s}{1-\omega} \tag{3.77c}$$

we have the difference equation,

$$\frac{U_{j-1}}{C_{j-1}} - \frac{U_j}{C_j} = \Pi_s \binom{s}{j} \frac{s[1-a^*(j\mu)] - j}{[1-a^*(j\mu)][s(1-\omega)-j]C_j}. \tag{3.78a}$$

where C_j is defined in (3.71).

Summing (3.78a) from $j = i+1$ to $s-1$, and noting $U_{s-1} = \Pi_{s-1} = \Pi_s/\omega$, we have

$$\frac{U_i}{C_i} = \Pi_s \sum_{j=i+1}^{s} \binom{s}{j} \frac{s[1-a^*(j\mu)] - j}{[1-a^*(j\mu)][s(1-\omega)-j]C_j}, \quad i = 0, 1, \cdots, s-1. \tag{3.78b}$$

Setting $i = 0$, and noting U_0 in (3.77c) and $C_0 = 1$, we obtain (3.79).

3.5 Renewal Input Single Server Models

3.5.1 Spectral Solution for GI/G/1

An exact analysis for GI/G/1 models has been given by using the *spectral solution*. Here, outline of the analysis and useful formulas are presented. For details, refer to [8, Vol.I, p.275].

For the rth call in a GI/G/1 system, denote the service time by X_r, the waiting time by W_r, and the interarrival time between the rth and $(r+1)$st call by Y_{r+1}, as shown in Figure 3.9. Then, we have the relation,

$$W_{r+1} = (W_r + X_r - Y_{r+1})^+ \tag{3.87}$$

where $(z)^+ = \max(0, z)$, which is equal to z if $z \geq 0$, or 0 if $z < 0$. Assuming the existence of a steady state as $r \to \infty$, and letting $W_r \to W$ and $(X_r - Y_{r+1}) \to U$, from (3.87) we have

$$W = (W + U)^+. \tag{3.88}$$

It is known that a steady state exists if and only if the server utilization $\rho = \lambda h < 1$ with an arrival rate λ and a mean service time h.

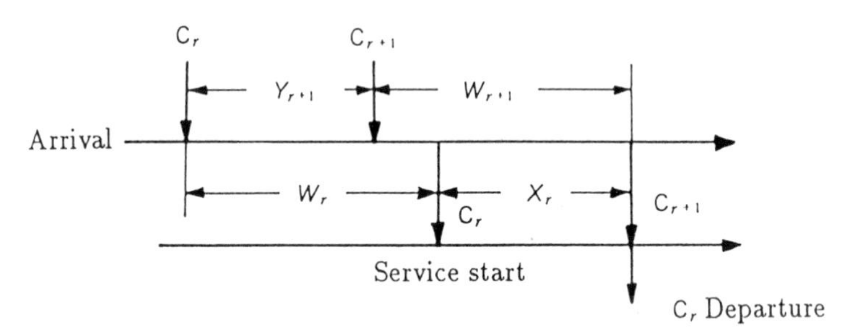

Fig.3.9 GI/G/1 model

Letting $u(t)$ be the density function of U, and $W(t)$ the waiting time distribution function, from (3.88) we get the *Lindley integral equation*,

$$W(t) = \int_{-\infty}^{t} W(t - v)u(v)dv, \ t \geq 0. \tag{3.89}$$

Introducing the auxiliary function $W_-(t)$ defined so that (3.89) also holds for $-\infty < t < 0$, we have the expression,

$$W(t) + W_-(t) = \int_{-\infty}^{t} W(t - v)u(v)dv, \ -\infty < t < \infty. \tag{3.90}$$

Since U is the difference between service time X and interarrival time Y, the LST of U is given by

$$u^*(\theta) = a^*(\theta)b^*(-\theta) \tag{3.91}$$

where $a^*(\theta)$ and $b^*(\theta)$ are the LSTs of X and Y, respectively. (See convolution formula in Appendix C [T5].) Denoting the Laplace transform (LT) of $W(t)$ by $W^*(\theta)$, transforming both sides of (3.90), and using (3.91), we have

$$W^*(\theta)[a^*(\theta)b^*(-\theta) - 1] = W_-^*(\theta). \tag{3.92}$$

Now, assume that [ ] in the left hand side of (3.92) has a spectral factorization of the form,

$$a^*(\theta)b^*(-\theta) - 1 = \frac{\Psi_+(\theta)}{\Psi_-(\theta)} \tag{3.93}$$

which satisfies the following condition:

(1) $\Psi_+(\theta)$ is analytic for $\text{Re}(\theta) > 0$, with no zeros in this half-plane, and

$$\lim_{|\theta|\to\infty} \frac{\Psi_+(\theta)}{\theta} = 1. \tag{3.94}$$

(2) $\Psi_-(\theta)$ is analytic for $\text{Re}(\theta) < D$, with no zeros in this half-plane, and

$$\lim_{|\theta|\to\infty} \frac{\Psi_-(\theta)}{\theta} = -1 \tag{3.95}$$

where D is a positive real number.

Substituting (3.93) into (3.92), we have

$$W^*(\theta)\,\Psi_+(\theta) = W_-^*(\theta)\,\Psi_-(\theta). \tag{3.96}$$

From Conditions (1) and (2) above, since both sides of (3.96) are analytic in the region $0 < \text{Re}(\theta) < D$, they should be equal to a constant, say K, from the *Liouville theorem* which states that if $f(z)$ is analytic and bounded for all finite values of z, then $f(z)$ is a constant. Since the LST of $W(t)$ is given by $w^*(\theta) = \theta W^*(\theta)$, from the *initial value theorem* (See Appendix C [T5].) and (3.94), we can determine the constant K as

$$\lim_{\theta\to\infty} \theta W^*(\theta) = W(0^+) = K.$$

Using the relation $w^*(\theta) \to 1$ as $\theta \to 0$, we have the probability of no waiting,

$$W(0^+) = \lim_{\theta\to 0} \frac{\Psi_+(\theta)}{\theta} \tag{3.97}$$

and the LST of the waiting time distribution,

$$w^*(\theta) = W(0^+)\frac{\theta}{\Psi_+(\theta)}. \tag{3.98}$$

Inverting $w^*(\theta)$ we can calculate the waiting time distribution, and differentiating $w^*(\theta)$ at the origin gives the moments of the waiting time. (See Appendix C [T5].) For example, we have the mean waiting time,

$$W = -\lim_{\theta\to 0} \frac{dw^*(\theta)}{d\theta}. \tag{3.99}$$

3.5.2 $H_2/G/1$

Consider the system for which the interarrival time is distributed with H_2 (hyper-exponential distribution) with the LST,

$$a^*(\theta) = \frac{k\lambda_1}{\theta + \lambda_1} + \frac{(1-k)\lambda_2}{\theta + \lambda_2}. \tag{3.100}$$

The parameters are determined in terms of the arrival rate λ, the SCV C_a^2, and skewness S_k (third central moment/variance$^{3/2}$) of the interarrival time distribution:

$$\begin{aligned} \lambda_1 &= \lambda \frac{C^{1/2} - S_k C_a^3 + 2}{3C_a^4 - 2S_k C_a^3 + 1} \\ C &\equiv (S_k^2 + 18)C_a^6 - 12S_k C_a^5 - 18C_a^4 + 8S_k C_a^3 + 6C_a^2 - 2 \\ k &= \frac{(C_a^2 - 1)\lambda_1^2}{(C_a^2 + 1)\lambda_1^2 - 2(2\lambda_1 - \lambda)\lambda} \\ \lambda_2 &= \frac{\lambda\lambda_1(1-k)}{\lambda_1 - k\lambda}. \end{aligned} \tag{3.101}$$

In particular, if only λ and C_a^2 are given, setting the *symmetrical condition*,

$$\frac{k}{\lambda_1} = \frac{1-k}{\lambda_2},$$

we have

$$k = \frac{1}{2}\left(1 + \sqrt{1 - \frac{2}{1 + C_a^2}}\right), \quad \lambda_1 = 2k\lambda, \quad \lambda_2 = 2(1-k)\lambda. \tag{3.102}$$

Now, we shall present the result for the $H_2/G/1$ by applying the spectral solution. According to (3.93), if we set

$$\begin{aligned} \Psi_+(\theta) &= \frac{y(\theta)b^*(\theta) - (\lambda_1 - \theta)(\lambda_2 - \theta)}{\theta_0 - \theta} \\ \Psi_-(\theta) &= \frac{(\lambda_1 - \theta)(\lambda_2 - \theta)}{\theta_0 - \theta} \end{aligned} \tag{3.103}$$

we can verify that (3.103) satisfy (3.94) and (3.95) with $D = \min(\lambda_1, \lambda_2, \mathrm{Re}(\theta_0))$, respectively. Here, θ_0 is a real root of the functional equation,

$$y(\theta)b^*(\theta) - (\lambda_1 - \theta)(\lambda_2 - \theta) = 0 \tag{3.104}$$

where

$$y(\theta) = k\lambda_1(\lambda_2 - \theta) + (1-k)\lambda_2(\lambda_1 - \theta).$$

We can calculate θ_0 by iterating

$$\theta = \frac{z(\theta) - \sqrt{z(\theta)^2 - 4\lambda_1\lambda_2[1 - b^*(\theta)]}}{2} \tag{3.105}$$

with the initial value $\theta^0 = \lambda$, where

$$z(\theta) \equiv \lambda_1 + \lambda_2 - [k\lambda_1 + (1-k)\lambda_2]b^*(\theta).$$

From (3.97) to (3.99), we have the following result:

$$W(0^+) = (1-\rho)\frac{\lambda_1\lambda_2}{\lambda\theta_0} \tag{3.106}$$

$$w^*(\theta) = \frac{W(0^+)\theta(\theta_0 - \theta)}{y(\theta)b^*(\theta) - (\theta - \lambda_1)(\theta - \lambda_2)} \tag{3.107}$$

$$W = W_M + \frac{[k\lambda_1 + (1-k)\lambda_2]\rho - \lambda}{\lambda_1\lambda_2(1-\rho)} + \frac{1}{\theta_0} \tag{3.108}$$

where W_M is the mean waiting time of the equivalent M/G/1 (with the same arrival rate), given from (3.20) by

$$W_M = \frac{\rho(1 + C_s^2)}{1(1-\rho)}h \tag{3.109}$$

with the SCV C_s^2 of the service time distribution.

[Example 3.9] Consider the $H_2/D/1$ system with $\lambda = 0.5$/sec, $C_a^2 = 10$, $h = 1$ sec and $C_s^2 = 0$.

First, assuming the symmetric condition, from (3.102) we have

$$k = 0.0477, \quad \lambda_1 = 0.0477, \quad \lambda_2 = 0.9523/\text{sec}.$$

We get $\theta_0 = 2565$/sec by iterating (3.105), the waiting probability $M(0) = 1 - W(0^+) = 0.8228$, and the mean waiting time $W = 2.3990$ sec, respectively, from (3.106) and (3.108). It should be noted that, in this case, the skewness of the interarrival time is implicitly determined as $S_k = 9.4552$.

Next, given the skewness S_k, numerical results are shown in Table 3.1, in which the parameters are determined by using (3.101). It will be recognized that the skewness has a significant effect on the result.

Table 3.1 Numerical Example for $H_2/D/1$

$\lambda = 0.5$/sec, $C_a^2 = 10$, $h = 1$ sec

S_k	k	λ_1	λ_2	$M(0)$	W[sec]
6	0.1203	0.0742	2.3182	0.9164	7.0795
8	0.0678	0.0564	1.1681	0.8758	3.7937
10	0.0424	0.0451	0.9029	0.8022	2.0802
12	0.0287	0.0375	0.7868	0.7380	1.4257

3.5.3 $E_k/G/1$

Consider a system having an interarrival time distributed with E_k (Erlangian Distribution) for which LST is given by

$$a^*(\theta) = \left(\frac{k\lambda}{\theta + k\lambda}\right)^k. \tag{3.110}$$

Applying the spectral solution in a similar manner as above, we have the following results:

$$W(0^+) = \frac{(k\lambda)^k(1-\rho)}{\lambda}\prod_{i=1}^{k-1}\frac{1}{\theta_i} \tag{3.111}$$

$$w^*(\theta) = \frac{\theta W(0^+)}{(k\lambda)^k b^*(\theta) - (k\lambda - \theta)^k}\prod_{i=1}^{k-1}(\theta_i - \theta) \tag{3.112}$$

$$W = W_M - \frac{1 - C_a{}^2}{2\rho(1-\rho)}h + \sum_{i=1}^{k-1}\frac{1}{\theta_i} \tag{3.113}$$

where θ_i, $i = 1, 2, \cdots, k-1$, are the roots with $\mathrm{Re}(\theta_i) > 0$ of the functional equation,

$$(k\lambda)^k b^*(\theta) - (k\lambda - \theta)^k = 0. \tag{3.114}$$

Setting $b^*(\theta) = \gamma \exp(j\omega)$ with $j = \sqrt{-1}$, these roots can be calculated by iterating

$$\begin{aligned} \mathrm{Re}(\theta_i) &= \lambda k \left[1 - \gamma^{1/k}\cos\left(\frac{\omega + 2\pi i}{k}\right)\right] \\ \mathrm{Im}(\theta_i) &= -\lambda k \gamma^{1/k}\sin\left(\frac{\omega + 2\pi i}{k}\right), \quad i = 1, 2, \cdots, k-1. \end{aligned} \tag{3.115}$$

If k is even, we have one real root and $(k/2 - 1)$ conjugate pairs; and if k is odd, $(k-1)$ conjugate pairs. Hence, $\prod \theta_i$ and $\sum 1/\theta_i$ become real.

It should be noted that (3.111) to (3.113) are valid only for an integer k, and the skewness of the interarrival time is implicitly determined. An extended version for a real number k is also available, given the first three moments. For details, refer to [14].

[Example 3.10] Consider an $E_5/H_2/1$ with $\lambda = 0.5$/sec, $C_a{}^2 = 1/k = 0.5$, $h = 1$ sec and $C_s{}^2 = 2$ (symmetric). From (3.115), we get

$$\mathrm{Re}(\theta_1) = \mathrm{Re}(\theta_4) = 2.0772, \quad -\mathrm{Im}(\theta_1) = \mathrm{Im}(\theta_4) = 1.9551$$

$$\mathrm{Re}(\theta_2) = \mathrm{Re}(\theta_3) = 4.0560, \quad -\mathrm{Im}(\theta_2) = \mathrm{Im}(\theta_3) = 1.0439 \text{ (1/sec)}.$$

Using $\sum 1/\theta_i = 0.9730\,\text{sec}$ in (3.113), we have $W = 0.8730\,\text{sec}$.

Figure 3.10 shows a numerical example of (3.108) for $C_a{}^2 > 1$, and (3.113) for $C_a{}^2 \leq 1$, with a service time of H_2 for $C_s{}^2 > 1$, and of E_k for $C_s{}^2 \leq 1$. (Symmetric H_2 is used both for interarrival and service time.) The effect of $C_a{}^2$ and $C_s{}^2$ on the mean waiting time may be seen from the figure.

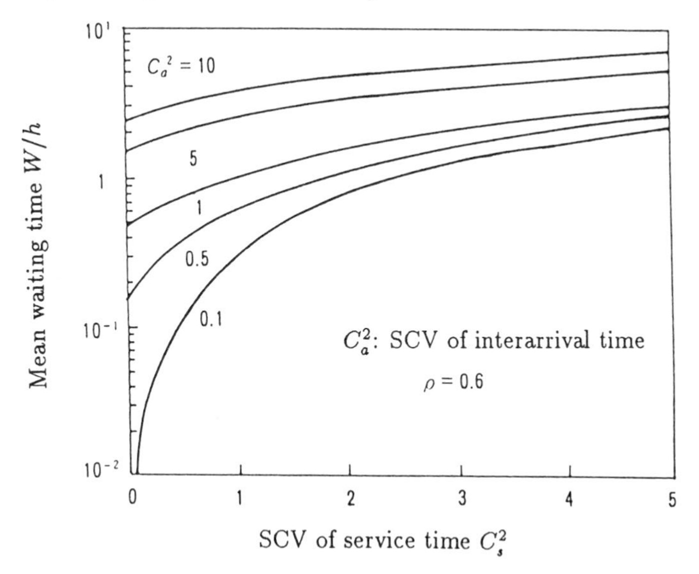

Fig.3.10 Mean waiting time of GI/G/1

Exercises

[1] Suppose that documents are transmitted by a facsimile machine. Assume that the documents arrive at random, of which 50 % is of fixed form with constant transmission (service) time 1 min, whereas the rest are exponentially distributed with the same mean.

(1) How many documents can be transmitted with mean waiting time no greater than 30 sec?

(2) Calculate the mean waiting time, if all documents are of the fixed form.

[2] Data packets with fixed length of 2400 bits originate at random at a rate of 25 packets/sec, and are transmitted via 2 lines of 32 kb/s with a sufficiently large buffer and the FIFO discipline.

(1) Calculate the mean response time and the probability that the response time exceeds 0.1 sec.

(2) Evaluate the above performance values when the line is replaced by a single 64 kb/s line.

[3] Suppose that computer jobs randomly arriving with mean interval 10 sec, are processed alternately by 2 CPUs, for which processing time is exponentially distributed with mean 15 sec. Obtain

(1) The mean waiting time and the probability that the waiting time exceeds 1 min, on an FIFO delay basis.

(2) The probability that the jobs are not processed by blocking, on a loss basis.

[4] Packetized voice with fixed length of 64 bytes originating at rate $\lambda = 21.96$ packets/sec, are transmitted via a line of speed 64 kb/s, with a sufficiently large buffer and FIFO discipline. The interarrival time is characterized by the SCV $C_a{}^2 = 18.095$ and skewness $S_k = 9.838$.

(1) Approximating the interarrival time by H_2 (asymmetric), estimate the mean waiting time for $H_2/D/1$ model.

(2) Using the symmetric type H_2, re-estimate the mean waiting time.

Chapter 4

MULTI-CLASS INPUT MODELS

This chapter deals with traffic models with multi-class input which appear in the ISDN (integrated services digital network) and LAN (local area network). These include batch arrival, priority queue, multi-dimensional, mixed loss and delay, multi-queue models, etc.

4.1 Batch Arrival Models

The *batch arrival model* in which a number of calls arrive at a time, appears in packet switching systems and facsimile communications. For example, in packet switching, a message is partitioned into a number of packets, which will be approximated as a batch arrival model by assuming packets for a message to arrive simultaneously at a time. In addition, there is the *batch service model* in which a number of calls are served in a batch. Typical examples of batch arrival models are analysed, in this section.

4.1.1 $\mathrm{M}^{[X]}/\mathrm{M}/s(0)$

Let X be a random variable representing the *batch size* which is defined by the number of calls simultaneously arriving in a batch. A loss system with batch size X, exponential batch interarrival time, exponential service time and s servers, is expressed as $\mathrm{M}^{[X]}/\mathrm{M}/s(0)$. If a batch finds all servers busy upon arrival, all calls in the batch are lost. If the number of idle servers is less than the batch size, as shown in Figure 4.1, there are two strategies for accepting calls in the batch:

(a) *Partial batch acceptance strategy* (PBAS): Only the same number of calls as that of idle servers, are served and the rest of the calls are lost.

(b) *Whole batch acceptance strategy* (WBAS): Unless all calls in a batch can be accepted, the batch is lost.

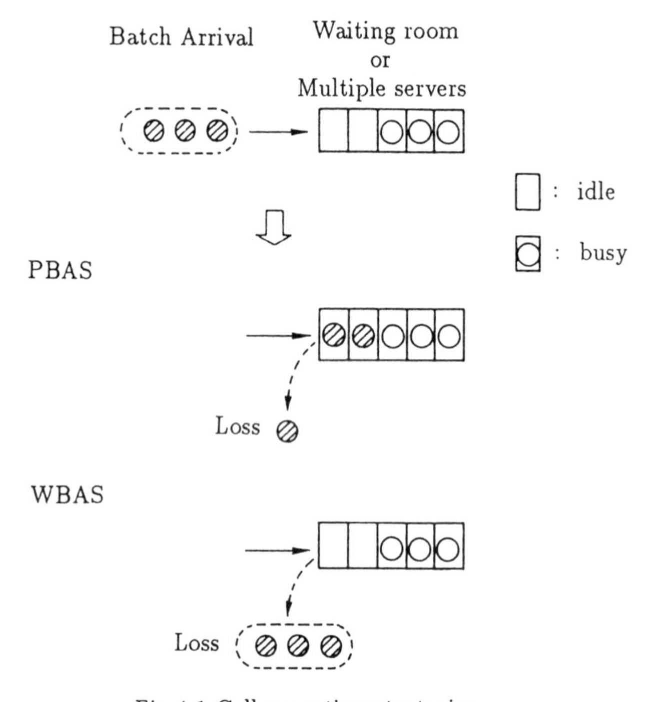

Fig.4.1 Call accepting strategies

In what follows, the PBAS is assumed in the steady state, and the following notation is used:

P_j : Probability that j calls exist at an arbitrary instant, $j = 0, 1, \cdots$.

$P(z)$: Probability generating function of P_j, $P(z) = \sum_{j=0}^{\infty} z^j P_j$. (4.1)

b_i : Probability that batch size is i, $b_i = P\{X = i\}$, $i = 1, 2, \cdots$.

$g(z)$: Probability generating function of b_i, $g(z) = \sum_{i=0}^{\infty} z^i b_i$. (4.2)

b : Mean batch size, $b = \sum_{i=1}^{\infty} i b_i$. (4.3)

$$\phi_i \ : \ \text{Probability of } X \geq i, \ \phi_i = \sum_{k=i}^{\infty} b_k, \quad i = 1, 2, \cdots \tag{4.4}$$

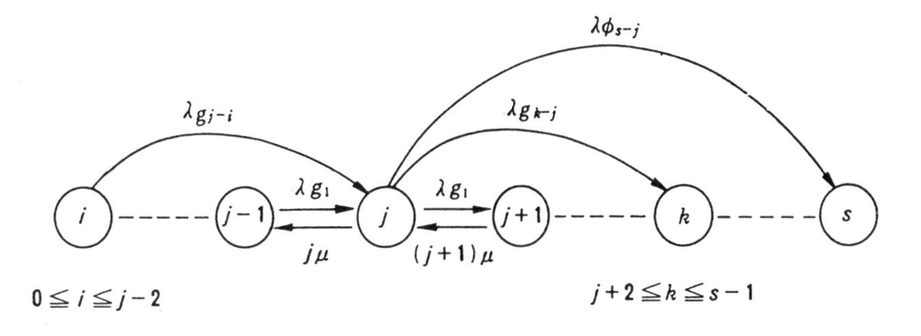

(a) $1 \leqq j \leqq s-1$

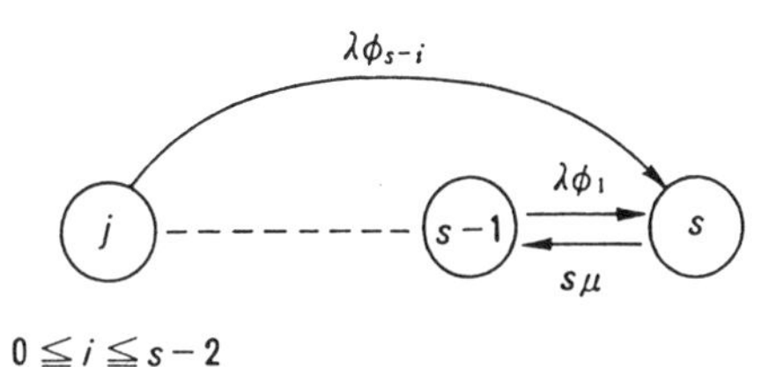

(b) $j = s$

Fig.4.2 State transition diagram for $M^{[X]}/M/s(0)$

If λ is the batch arrival rate, and μ^{-1} the mean service time of individual calls, the state transition diagram is given as shown in Figure 4.2. From the rate-in=rate-out, we have the steady state equation,

$$\lambda \sum_{i=0}^{j-1} P_i b_{j-i} + (j+1)\mu P_{j+1} = (\lambda + k\mu) P_j, \quad j = 0, 1, \cdots, s-1 \tag{4.5}$$

$$\lambda \sum_{i=0}^{s-1} P_i \phi_{s-i} = s\mu P_s \tag{4.6}$$

and the normalization condition,

$$\sum_{j=0}^{s} P_j = 1. \tag{4.7}$$

Summing (4.5) from $j = 0$ to $(j-1)$, leads to the recurrence formula,

$$P_j = \frac{\lambda}{j\mu}\sum_{i=0}^{j-1} P_i \phi_{j-i}, \quad j = 1, 2, \cdots, s. \tag{4.8}$$

Using this together with (4.7), we can calculate P_j recursively.

The blocking probability is given by

$$B = \frac{a - a_c}{a} = 1 - \frac{1}{a}\sum_{j=1}^{s} jP_j \tag{4.9}$$

where $a = \lambda bh$ is the offered load, $h = \mu^{-1}$ the individual mean service time, and $a_c = \sum_{j=1}^{s} jP_j$ the carried load. Note that $B = P_s$ does not hold for batch arrival models.

If the batch size X is geometrically distributed with

$$b_i = pq^{i-1}, \quad i = 1, 2, \cdots \tag{4.10}$$

where $q = 1 - p$. Hence, $\phi_i = q^{i-1}$ and (4.8) is rewritten as

$$P_j = \frac{\lambda}{j\mu}\sum_{i=0}^{j-1} P_i q^{j-i-1}, \quad j = 1, 2, \cdots, s. \tag{4.11}$$

In this case, denoting the blocking probability with s trunks by B_s, we have the recurrence formula,

$$B_s = \frac{(pa + qs)B_{s-1}}{s + paB_{s-1}}, \quad s = 1, 2, \cdots; \; B_0 = 1. \tag{4.12}$$

If $p = 1$, the model reduces to a non-batch arrival (Poisson input) model, and (4.12) coincides with (2.26a) for the Erlang B formula.

[Example 4.1] Consider a facsimile transmission system in which documents are transmitted simultaneously to a number of destinations.

Assume that documents arrive at random at rate λ, the transmission time is exponentially distributed with mean h, the number X of destinations is geometrically distributed with mean b, and s circuits transmit the facsimile signals on a loss basis. Therefore, the system is modeled by $\mathrm{M}^{[X]}/\mathrm{M}/s(0)$. For example, if $\lambda = 12/\mathrm{min}$, $h = 10\,\mathrm{sec}$ and $b = 5$, we have

$$a = \lambda bh = 12 \times \frac{10}{60} \times 5 = 10\,\mathrm{erl}$$

$$p = 1/b = 0.2, \quad q = 1 - p = 0.8. \quad \text{(See Table C.1 in Appendix C.)}$$

Hence from (4.12), we obtain the blocking probability $B = 0.1644$.

If we apply the M/M/s(0) model with the same offered load, the blocking probability becomes $B = 0.00846$ which is very much underestimated.

Figure 4.3 shows calculated results for various offered loads with the same transmission time and batch size as above. The blocking probability is finite even if $a \to 0$, because of the batch arrival. As a comparison, the results for M/M/s(0) (non-batch) are also shown by the broken lines.

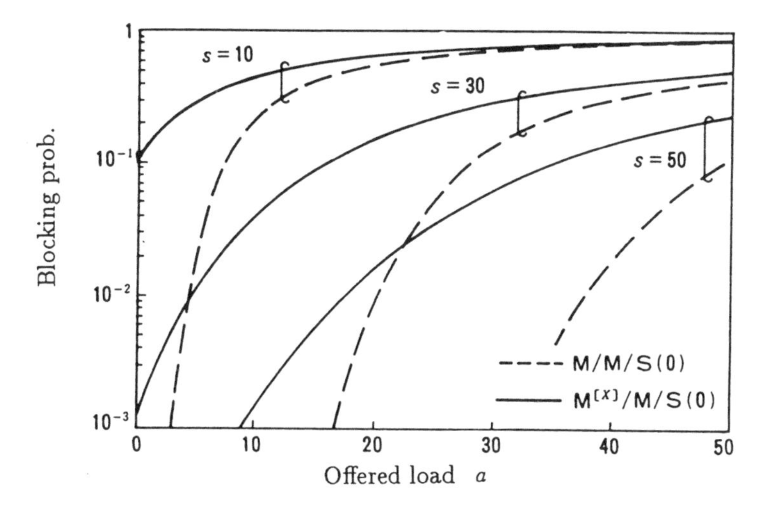

Fig.4.3 Blocking probability of batch arrival model

4.1.2 $M^{[X]}/G/1$

Consider the batch arrival delay system with a general service time distribution, $M^{[X]}/G/1$. This model is an extension of M/G/1 described in Section 3.2, in which calls arrive in batches. Using the same notation as before, a steady state exists if and only if $a = \lambda bh < 1$.

Mark an arbitrary call in a batch, and refer to as the test call. Then, the waiting time of the test call consists of the following two parts:

(1) Waiting time W_1 until the first call in the batch, to which the test call belongs, enters service. Batches are assumed to be served in FIFO (first-in first-out).

(2) Waiting time W_2 due to calls served before the test call in the batch. Calls in a batch are assumed to be served in RSO (random service order).

A batch is regarded as a hypothetical call having the composite service time equal to the sum of service times of calls belonging to the batch. Since the batches (hypothetical calls) arrive as a Poisson process, the model reduces to M/G/1 with the composite service time. Hence, letting h_b and $C_b{}^2$ be the mean and SCV of the composite service time, respectively, from (3.20) the mean of W_1 is given by

$$\overline{W}_1 = \frac{a}{1-a}\frac{1+C_b{}^2}{2}h_b. \tag{4.13}$$

Denote the conditional distribution function of the composite service time T_b, given the batch size $X = i$, by $B_i(t) = P\{T_b \leq t \mid X = i\}$. Then, the distribution function $B_b(t)$ of T_b is given by (from the total probability theorem in Appendix C [T2].)

$$B_b(t) = P\{T_b \leq t\} = \sum_{i=0}^{\infty} B_i(t)b_i. \tag{4.13a}$$

Note that since T_b is i hold convolution of the service time of an individual call, the LST $b_i^*(\theta)$ of $B_i(t)$ is given by $b_i^*(\theta) = [b^*(\theta)]^i$, where $b^*(\theta)$ is the LST of the individual service time. Hence, we have the LST of $B_b(t)$,

$$b_b^*(\theta) = \sum_{i=0}^{\infty}[b^*(\theta)]^i b_i = g(b^*(\theta)). \tag{4.14}$$

Thus, the mean h_b and second moment $h_b^{(2)}$ of the composite service time are given by

$$\begin{aligned} h_b &= -\lim_{\theta\to 0}\frac{db_b^*(\theta)}{d\theta} = bh \\ h_b^{(2)} &= \lim_{\theta\to 0}\frac{d^2b_b^*(\theta)}{d\theta^2} = bh^{(2)} + (b^{(2)} - b)h^2 \end{aligned} \tag{4.15}$$

where h and $h^{(2)}$ are the mean and second moment of the individual service time, respectively. Noting the relations,

$$\begin{aligned} a &= \lambda h_b = \lambda bh \\ C_b{}^2 &= \frac{h_b^{(2)} - h_b^2}{h_b^2} = \frac{C_s{}^2}{b} + \frac{b^{(2)}}{b^2} - 1 \end{aligned} \tag{4.15a}$$

where $C_s{}^2$ is the coefficient of variation of the individual service time. Using (4.15a) in (4.13), we obtain the mean of W_1,

$$\overline{W}_1 = \frac{\lambda h_b^{(2)}}{2(1-a)}. \tag{4.16}$$

To evaluate W_2, first observe that the probability r_k that a call is served in the kth order in the batch, is given by

$$r_k = \frac{1}{b}\sum_{i=k}^{\infty} b_i, \quad k = 1, 2, \cdots \tag{4.17}$$

which is derived as follows:

Denoting the batch size of the ith batch by X_i, the probability $x_i(k)$ that a call is served in the kth order in the ith batch, is clearly given by

$$x_i(k) = \begin{cases} 1, & X_i \geq k \\ 0, & \text{otherwise.} \end{cases} \tag{4.17a}$$

Taking n batchès from the beginning, the rate at which a call is served in the kth order, is given by

$$\frac{\text{No. of calls served in } k\text{th order}}{\text{Total No. of calls in } n \text{ batches}} = \frac{\sum_{i=1}^{n} x_i(k)}{\sum_{i=1}^{n} X_i}. \tag{4.17b}$$

Since r_k is the limit of the rate, (See law of large number in Appendix C [T1].) dividing the numerator and denominator by n, and taking the limit we have

$$r_k = \lim_{n\to\infty} \frac{\sum_{i=1}^{n} x_i(k)/n}{\sum_{i=1}^{n} X_i/n} = \frac{P\{X_i \geq k\}}{E\{X_i\}}, \quad k = 1, 2, \cdots. \tag{4.17c}$$

from which (4.17) follows.

Now, the LST $w_2^*(\theta)$ of the distribution function of W_2, similarly to (4.14), is given by

$$w_2^*(\theta) = \sum_{k=1}^{\infty} [b^*(\theta)]^{k-1} r_k = \frac{1 - g(b^*(\theta))}{b[1 - b^*(\theta)]} \tag{4.17d}$$

where we note that the waiting time is 0 when $k = 1$. Hence, the mean of W_2 is given by

$$\overline{W}_2 = -\lim_{\theta\to 0} \frac{dw_2^*(\theta)}{d\theta} = \frac{b^{(2)} - b}{2b} h. \tag{4.18}$$

Since the test call has been chosen arbitrarily, the mean overall waiting time for an arbitrary call is given by

$$\begin{aligned} W &= \overline{W}_1 + \overline{W}_2 = \frac{\lambda h_b^{(2)}}{2(1-a)} + \frac{h}{2}\left(\frac{b^{(2)}}{b} - 1\right) \\ &= \frac{1}{2(1-a)}\left[\lambda b h^{(2)} + \left(\frac{b^{(2)}}{b} - 1\right) h\right]. \end{aligned} \tag{4.19}$$

[Example 4.2] Consider a data transmission system in which messages arrive at random at rate λ, and are transmitted on FIFO delay basis with infinite buffer via a single link with speed v [b/s], packetized into fixed size packets of l [bit].

Assuming the length of message is exponentially distributed with mean m [bit], the system is approximated by $M^{[X]}/D/1$ with X geometrically distributed with mean $b = m/l$. For example, if $\lambda = 12$ messages/sec, $m = 4000$ bits, $l = 2000$ bits, and $v = 64$ kb/s, we have

$$h = l/v = 31.25\,\text{ms}, \quad b = m/l = 2, \quad b^{(2)} = b(2b-1) = 6$$

$$a = \lambda b h = 0.75\,\text{erl}.$$

With $h^{(2)} = h^2$ for fixed packet size, from (4.16), (4.18) and (4.19) we have

$$\overline{W}_1 = 0.1406\,\text{sec}, \quad \overline{W}_2 = 0.03125\,\text{sec}, \quad W = 0.1719\,\text{sec}.$$

Figure 4.4 shows a numerical example of the normalized mean waiting time W/h for various offered loads and geometric batch sizes, where fixed and exponential service (transmission) times are compared, with $h^{(2)} = 2h^2$ for exponential service time.

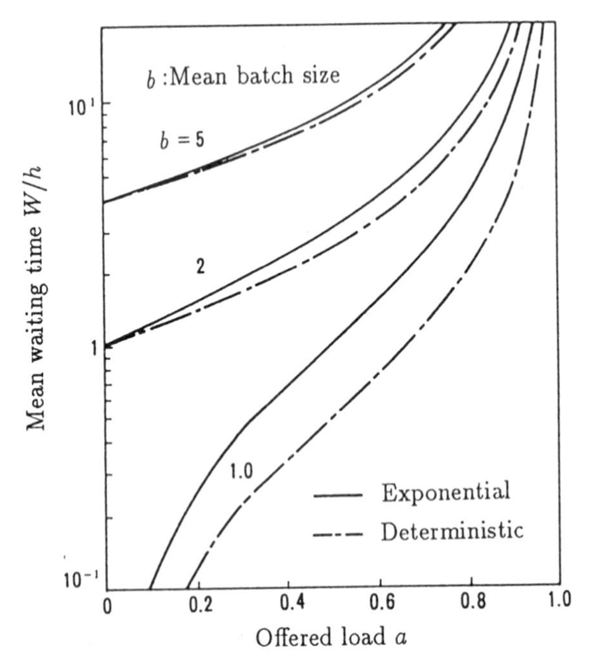

Fig.4.4 Mean waiting time of $M^{[X]}/G/1$

4.1.3 $GI^{[X]}/G/1$

This is a general batch arrival delay system in which the batch interarrival time and the individual service time are arbitrarily distributed. For this model, a similar approach as above can be applied except that the formula for GI/G/1 is needed instead of (4.13) for evaluating $\overline{W}_1$. If the batch interarrival time is distributed with H_2 or E_k, (3.108) or (3.113) may be used.

For general GI/G/1, the following approximation by *Kraemer and Langenbach-Belz formula* has been proposed [15]:

$$W = \frac{a}{1-a} \cdot \frac{(C_a{}^2 + C_s{}^2)J}{2} h \tag{4.20}$$

where $C_a{}^2$ and $C_s{}^2$ are the SCVs of the interarrival time and service time, respectively, and J is given by

$$J = \begin{cases} \exp\left[-\dfrac{2(1-a)(1-C_a{}^2)^2}{3a(C_a{}^2 + C_s{}^2)}\right], & C_a{}^2 \le 1 \\ \exp\left[-(1-a)\dfrac{C_a{}^2 - 1}{C_a{}^2 + 4C_s{}^2}\right], & C_a{}^2 > 1. \end{cases} \tag{4.21}$$

This formula coincides with Pollaczek-Khintchine formula in (3.20) with $C_a{}^2 = 1$ for Poisson input.

Using $h_b = bh$ for h, and λbh for a, we can approximate the mean waiting time $\overline{W}_1$. Since the mean delay $\overline{W}_2$ in a batch is independent of the batch arrival process and given by (4.18), the overall mean waiting time W is calculated from (4.19).

[**Example 4.3**] Suppose that in Example 4.2, the message interarrival time is distributed with H_2 (hyper-exponential distribution) with $C_a{}^2 = 10$.

With geometric batch size with mean $b = 2$ and $C_s{}^2 = 0$ for fixed packet size, from (4.15a) we have

$$C_b{}^2 = 1 + \frac{C_s{}^2 - 1}{b} = 0.5.$$

Using (3.108) instead of (4.13), for $H_2^{[X]}/E_k/1$ with $k = 1/C_b{}^2 = 2$ and mean service time $h_b = bh = 0.0625\,\text{sec}$, we have

$$\overline{W}_1 = 0.9013\,\text{sec}, \quad \overline{W}_2 = 0.03125\,\text{sec}, \quad W = 0.9317\,\text{sec}$$

where the symmetric H_2 is applied. If (4.20) is used instead, we have $\overline{W}_1 = 0.9844\,\text{sec}$ and $W = 1.0156\,\text{sec}$, which provides a fairly good approximation. It should be noted, however, that if the skewness of the H_2 were given, (4.20) would no longer apply. (See Example 3.9.)

4.2 Priority Models

In practice, priority is sometimes given to a certain class of calls to improve the grade of service (GOS) over other classes. For example, in a computer system handling batch and time sharing service (TSS) jobs, the TSS jobs are required to have a higher priority, for providing interactive processing. Similar models also appear in electronic switching and computer systems, and are referred to as *priority models.* There are two kinds of priorities: *external priority* inherently associated with the call before entering the system, and *internal priority* assigned after entering the system according to the system situation.

This section describes typical examples with external priority [16].

4.2.1 Non-Preemptive Priority Model

In this model, when all servers are busy, every time a call in service terminates, the waiting call with the highest priority enters service. Since a higher priority call does not interrupt a lower priority call in service, it is called a *non-preemptive priority model.*

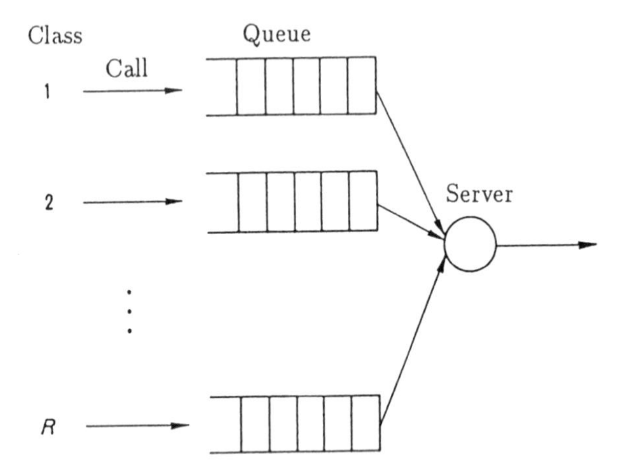

Fig.4.5 Priority Queueing Model

Consider the single server system shown in Figure 4.5, in which calls of class i, $i = 1, 2, \cdots, R$, arrive at random at rate λ_i with the service time arbitrarily distributed with mean h_i and second moment $h_i^{(2)}$. A higher priority class is given a smaller i. Let $\rho_i = \lambda_i h_i$, L_i, and W_i be the offered load, the mean number of waiting calls, and the mean waiting time, respectively, for class i calls. Assume FIFO with infinite buffer in the same class and the existence of a steady state.

If an arbitrary call in class i is chosen as the test call, its mean waiting time consists of the following three components:

(1) The mean time W_0 until a call terminates, if the call is in service.

Let r_i be the probability that a call being served is of class i. Then, this is equal to the traffic load of that class, *i.e.* $r_i = \rho_i$. (See Property (3) in Subsection 1.2.1.) Since the mean residual service time for class i calls is given from (3.9) by $h_i^{(2)}/(2h_i)$, we have

$$W_0 = \sum_{i=1}^{R} r_i \frac{h_i^{(2)}}{2h_i} = \frac{1}{2}\sum_{i=1}^{R} \lambda_i h_i^{(2)}. \tag{4.22}$$

(2) The mean time A_i until class i, or higher priority calls, waiting ahead of the test call, have been served.

From the Little formula, we have

$$A_i = \sum_{j=1}^{i} h_j L_j = \sum_{j=1}^{i} \rho_j W_j. \tag{4.22a}$$

(3) The mean time B_i due to the higher priority calls which arrive while the test call is waiting and are served before the test call.

Noting that the mean number of class j calls, $j = 1, 2, \cdots, i-1$, arriving in the mean waiting time W_i, is $\lambda_j W_i$, each having the mean service time h_j, and $B_1 = 0$, we have

$$B_i = \sum_{j=1}^{i-1} W_i \lambda_j h_j = W_i \sum_{j=1}^{i-1} \rho_j. \tag{4.22b}$$

Since the mean waiting time W_i for class i is the sum of the above three components, it follows that

$$W_i = W_0 + \sum_{j=1}^{i} \rho_j W_j + W_i \sum_{j=1}^{i-1} \rho_j. \tag{4.23}$$

Solving (4.23) recursively, yields the mean waiting time for class i calls,

$$W_i = \frac{\sum_{i=1}^{R} \lambda_i h_i^{(2)}}{2\left(1 - \sum_{j=1}^{i-1} \rho_j\right)\left(1 - \sum_{j=1}^{i} \rho_j\right)}. \tag{4.24}$$

The mean system (sojourn) time T_i is given by

$$T_i = W_i + h_i. \tag{4.25}$$

4.2.2 Preemptive Priority Model

A model is called a *preemptive priority model* if an incoming higher priority call, finding a lower priority call being served, is served by interrupting the latter call. The following classification is made according to the method for re-starting the interrupted service:

(a) *Resume*: The residual service time is continued from the interrupted point.

(b) *Repeat-Identical*: The identical service time is repeated from the beginning.

(c) *Repeat-Different*: A different service time is repeated from the beginning.

For exponential service time, from the Markov property, (a) and (b) are the same condition, and a simplified solution is obtained, which will be presented below. Such elegant results have not been obtained for other service time distributions.

Let us analyze the M/M/1 type preemptive priority model using the same assumption and notation as before, except for the preemption. The mean system (waiting + service) time of the test call of class i, consists of the following two components:

(1) The mean waiting time W_i from the test call arrival until entering service.

Since the test call can preempt lower priority calls, if any, it is no need to account for the traffic loads of class $i+1,\ i+2, \cdots,\ R$. Since higher priority calls are served before the test call, any way, the preemption has no consequence on the waiting time of the test call. Hence, by neglecting the traffic loads of the lower priority calls in (4.24) for non-preemptive priority model, we have

$$W_i = \frac{\sum_{j=1}^{i} \lambda_j h_j^{(2)}}{2\left(1 - \sum_{j=1}^{i-1} \rho_j\right)\left(1 - \sum_{j=1}^{i} \rho_j\right)}. \tag{4.25a}$$

(2) The mean *completion time* C_i from the test call entering service until eventually leaving the system, suffering from possible preemptions by higher priority calls. This is shown in Figure 4.6.

During time C_i, $\lambda_j C_i$ calls of class j, $j = 1, 2, \cdots, i-1$, arrive, and each occupies the server for h_j on average. Hence, we have

$$C_i = h_i + C_i \sum_{j=1}^{i-1} \lambda_j h_j \tag{4.25b}$$

solving which gives the mean completion time,

$$C_i = \frac{h_i}{1 - \sum_{j=1}^{i-1} \rho_j}, \quad i = 1, 2, \cdots, R. \tag{4.26}$$

This may be interpreted as the mean service time h_i being extended to C_i due to preemptions by higher priority calls.

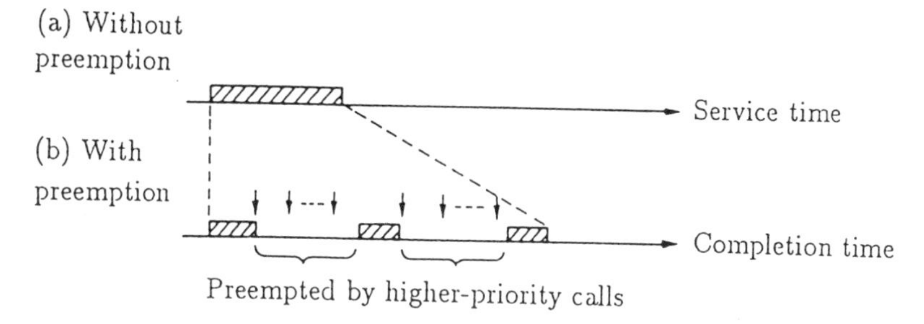

Fig.4.6 Completion time

Now, the mean system time of class i calls is given by $T_i = W_i + C_i$, or

$$T_i = \frac{1}{1 - \sum_{j=1}^{i-1} \rho_j} \left[\frac{\sum_{j=1}^{i} \lambda_j h_j^{(2)}}{2\left(1 - \sum_{j=1}^{i} \rho_j\right)} + h_i \right]. \tag{4.27}$$

The effective mean waiting time (including preempted time) is given by

$$W_i = T_i - h_i. \tag{4.28}$$

[Example 4.4] Consider a priority model with $R = 3$ and identical traffic conditions, $\lambda_i = \lambda$, $h_i = h$ and $h_i^{(2)} = h^2$, $i = 1, 2, 3$ (fixed service time).

If $\lambda = 0.2$/sec, $h = 1$ sec, from (4.24) and (4.28) we have

Non–preemptive priority : $W_1 = 0.375$, $W_2 = 0.625$, $W_3 = 1.250$ sec

Preemptive priority : $W_1 = 0.125$, $W_2 = 0.667$, $W_3 = 1.917$ sec.

Figure 4.7 shows a numerical example for various values of total offered load $\rho = \lambda h$.

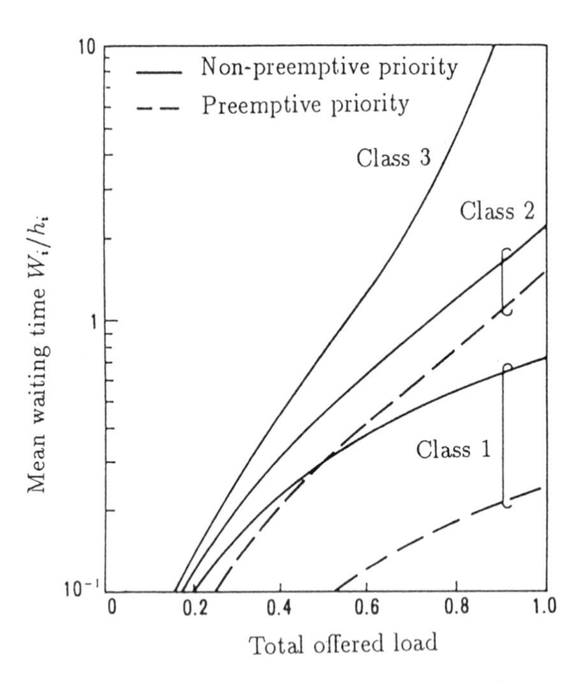

Fig.4.7 Mean waiting time for priority models

4.2.3 Work Conservation Law

Let us calculate the weighted total waiting time for the non-preemptive priority system. Summing (4.24) weighted by the offered load ρ_i, we have

$$\sum_{i=1}^{R} \rho_i W_i = \frac{\rho_0 W_0}{1-\rho_0} \tag{4.29}$$

where $\rho_0 = \rho_1 + \rho_2 + \cdots + \rho_R$ is the total offered load, and W_0 is the mean residual service time given in (4.22). Equation (4.29) implies that the weighted total waiting time is constant, is determined by the total offered load ρ_0 and the mean residual service time W_0, and is independent of priority assignment among classes. This is called the *M/G/1 work conservation law.*

It is known that a similar law holds for a more general work conservative model. A system is said to be *work conservative*, if the following condition is satisfied:

(1) The server continues handling calls without rest as long as any call exists in the system.

(2) The service time of a call is invariant once the call is accepted in the system: for example, no calls can leave the system in the way of service.

A preemptive repeat priority model is not work conservative since the repeated service caused by preemption creates an additional service time.

The *unfinished work* is defined as the total service time needed to finish at a certain time instant. Letting U be the mean unfinished work for a work conservative G/G/1 system, we have the *general work conservation law* [8, Vol.II, p.113]

$$\sum_{i=1}^{R} \rho_i W_i = U - W_0. \tag{4.29a}$$

For Poisson input, from the PASTA, the mean unfinished work U is equal to the mean waiting time W. Aggregating all classes into one class to construct the equivalent M/G/1 system, and noting the relation in (3.18a) for W_0, we have

$$U = W = \frac{W_0}{1 - \rho_0}. \tag{4.29b}$$

Substituting (4.29b) into (4.29a), we obtain (4.29) as a special case.

[Example 4.5] Consider a computer system which processes TSS (time sharing service), local batch and remote batch jobs, with priority classes $i = 1, 2$ and 3, respectively.

Assume that respective class jobs arrive at random at $\lambda_1 = 0.5$, $\lambda_2 = 0.2$ and $\lambda_3 = 0.05$ jobs/sec, and the means and second moments of service times are

$$h_1 = 1, \quad h_2 = 2, \quad h_3 = 3 \text{ sec}; \quad h_1^{(2)} = 5, \quad h_2^{(2)} = 8, \quad h_3^{(2)} = 9 \text{ sec}^2.$$

For the non-preemptive priority model, from (4.24) we have

$$W_1 = 3.75, \quad W_2 = 12.50, \quad W_3 = 41.67 \text{ sec},$$

and the left hand side of (4.29),

$$\sum_{i=1}^{3} \rho_i W_i = 10.625 \text{ sec}.$$

Using $\rho_0 = 0.85$ and $W_0 = 1.875$ sec, we have the right hand side of (4.29),

$$\frac{\rho_0 W_0}{1 - \rho_0} = 10.625 \text{ sec}.$$

Thus, the work conservation law holds.

If the preemptive priority model is assumed, from (4.28), we have

$$W_1 = 2.50, \quad W_2 = 13.00, \quad W_3 = 48.67 \text{ sec}; \quad \sum_{i=1}^{3} \rho_i W_i = 11.151 \text{ sec}.$$

Since this model is not work conservative, the work conservation law does not hold.

4.3 Multi-Dimensional Traffic Models

4.3.1 Multi-Dimensional Traffic

The *multi-dimensional traffic* is defined here as calls with different characteristics sharing common resources. Examples include mixed voice and data traffic with different transmission speed (or frequency band width) sharing a common group of transmission channels. Let us consider a multi-dimensional loss system to which low-speed (data) and high-speed (voice) calls are offered [17].

It is assumed that a low-speed call requires a unit channel (time slot), while a high-speed call requires m unit channels. In what follows, by channel we mean a unit channel. Consider the loss system in which low-speed and high-speed calls with Poisson arrival rates λ_1 and λ_2, and exponential service times of mean μ_1^{-1} and μ_2^{-1}, respectively, are offered to n channels as shown in Figure 4.8. It is assumed that, upon arrival, a low-speed call is blocked when all channels are busy, whereas a high-speed call is blocked unless at least m channels are available.

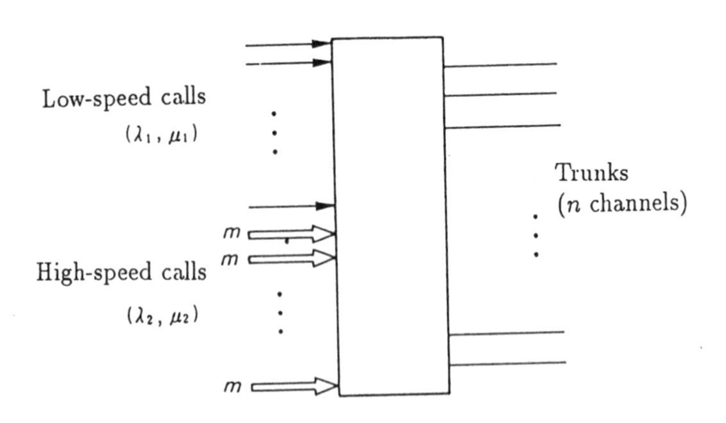

Fig.4.8 Multi-dimensional traffic model

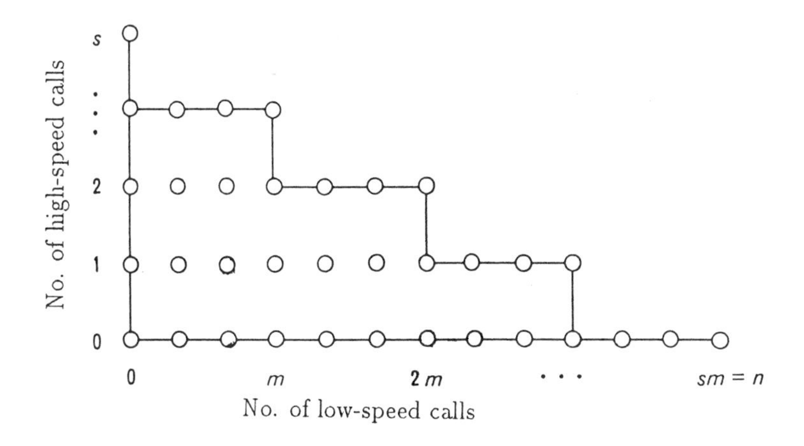

Fig.4.9 State space of Multi-dimensional model

Let P_{ij} be the joint probability that i low-speed calls and j high-speed calls exist in the steady state. Then, from Figure 4.9, we have the steady state equation,

$$\begin{aligned}(\lambda_1 + \lambda_2 + i\mu_1 + j\mu_2)P_{ij} &= \lambda_1 P_{i-1,j} + \lambda_2 P_{i,j-1} + (i+1)\mu_1 P_{i+1,j} \\ &\quad + (j+1)\mu_2 P_{i,j+1}, \quad 0 \leq i + mj \leq n - m\end{aligned} \tag{4.30}$$

$$(\lambda_1 + i\mu_1 + j\mu_2)P_{ij} = \lambda_1 P_{i-1,j} + \lambda_2 P_{i,j-1} + (i+1)\mu_1 P_{i+1,j}, \quad n - m < i + mj < n \tag{4.31}$$

$$(i\mu_1 + j\mu_2)P_{ij} = \lambda_1 P_{i-1,j} + \lambda_2 P_{i,j-1}, \quad i + mj = n \tag{4.32}$$

where $P_{ij} = 0$ for $i, j < 0$.

Denoting the individual traffic loads by $a_i = \lambda_i/\mu_i$, $i = 1, 2$, it can be shown that the *product form solution*,

$$P_{ij} = \frac{a_1{}^i}{i!}\frac{a_2{}^j}{j!}P_{00} \tag{4.33}$$

satisfies (4.30) to (4.32). From the normalization condition, P_{00} is determined as

$$P_{00} = \left(\sum_{j=0}^{s}\sum_{i=0}^{n-mj}\frac{a_1{}^i}{i!}\frac{a_2{}^j}{j!}\right)^{-1}. \tag{4.33a}$$

where $s = [n/m]$ with $[\,\cdot\,]$ representing the maximum integer involved.

Using (4.33), the blocking probabilities B_1 and B_2 for low-speed and high-speed calls, respectively, are given by

$$B_1 = \sum_{j=0}^{s} P_{n-mj,j} = P_{00} \sum_{j=0}^{s} \frac{a_1^{n-mj}}{(n-mj)!} \frac{a_2^{j}}{j!} \tag{4.34}$$

$$\begin{aligned} B_2 &= \sum_{i=0}^{k} P_{is} + \sum_{j=0}^{s-1} \sum_{i=n-mj-m+1}^{n-mj} P_{ij} \\ &= P_{00} \left(\frac{a_2^{2}}{s!} \sum_{i=0}^{k} \frac{a_1^{i}}{i!} + \sum_{j=0}^{s-1} \sum_{i=n-mj-m+1}^{n-mj} \frac{a_1^{i}}{i!} \frac{a_2^{j}}{j!} \right). \end{aligned} \tag{4.35}$$

where $k = n \pmod{m}$ represents the residue divided by m.

The channel efficiency is given by

$$\begin{aligned} \eta &= \frac{1}{n} [a_1(1 - B_1) + a_2 m(1 - B_2)] \\ &= \frac{A(1 - B)}{n} \end{aligned} \tag{4.36}$$

where A is the *total equivalent offered load*, and B the *average blocking probability* given by

$$\begin{aligned} A &= a_1 + m a_2 \\ B &= \frac{a_1 B_1 + m a_2 B_2}{A}. \end{aligned} \tag{4.36a}$$

[Example 4.6] Consider an ISDN system which integrates data and voice calls by circuit switching.

Assume that the data and voice calls arrive at random at rates $\lambda_1 = 15/\text{min}$, $\lambda_2 = 0.1/\text{min}$, and are transmitted at the speeds of 6.4 kb/s and 64 kb/s with holding times $\mu_1^{-1} = 0.2\,\text{min}$ and $\mu_2^{-1} = 2\,\text{min}$, respectively. Then, we have

$$a_1 = 1.5 \times 0.2 = 3\,\text{erl}, \quad a_2 = 0.1 \times 2 = 0.2\,\text{erl}, \quad m = 64/6.4 = 10.$$

From (4.34) to (4.36), we obtain

$$B_1 = 0.00276, \quad B_2 = 0.04378, \quad B = 0.01917.$$

Figure 4.10 shows the effect of the speed ratio m. The fluctuating characteristic appearing in the figure is called the *fraction channel effect*. This results from the fact that high-speed traffic is blocked unless a set of m channels is available, while the low-speed traffic can be served if a channel is available.

It is also seen from the figure that the blocking probability for the high-speed traffic is greater than for low-speed traffic, which results in an unbalanced GOS for different classes of traffic. Therefore, careful design is needed for systems integrating heterogeneous calls with different characteristics. To solve such problems, the trunk reservation scheme is used, which will be described next.

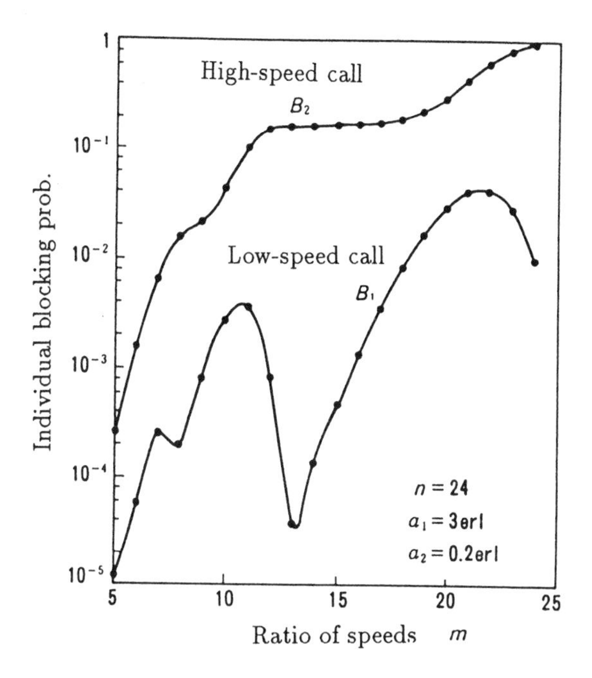

Fig.4.10 Blocking probabilities of multi-dimensional traffic model

4.3.2 Trunk Reservation System

The *trunk reservation system* is used to guarantee the GOS for preferential traffic, by reserving a certain number of trunks for that traffic. A numerical analysis is presented here for delay systems using a trunk reservation scheme with two classes of traffic: priority and ordinary traffic [18].

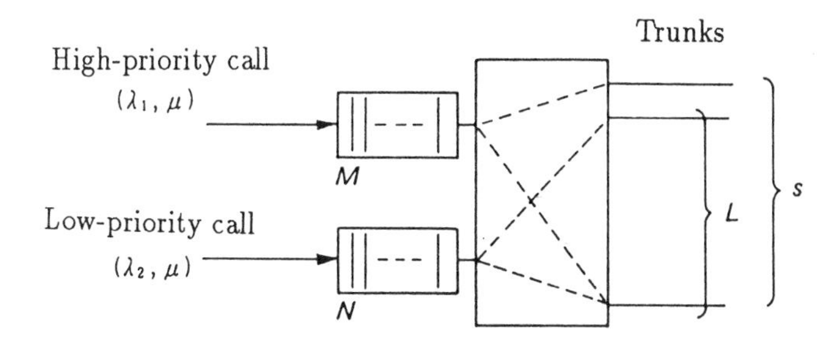

Fig.4.11 Delay-delay trunk reservation system

Consider the system shown in Figure 4.11, in which priority and ordinary calls arrive at Poisson rates λ_1 and λ_2, respectively, the service time is exponentially distributed with the same mean μ^{-1}, and an ordinary incoming call is rejected if the number of idle servers is no greater than R. The number of servers is s, and the capacities of the waiting rooms for the priority and ordinary traffic are M and N, respectively, with each class being served in FIFO discipline. We call this the delay-delay trunk reservation system, since both classes are served on a delay basis. By letting $M = 0$ and/or $N = 0$, we have loss-loss, loss-delay and delay-loss systems. If we set $R = 0$, the system reduces to the non-preemptive priority model described in Subsection 4.3.1.

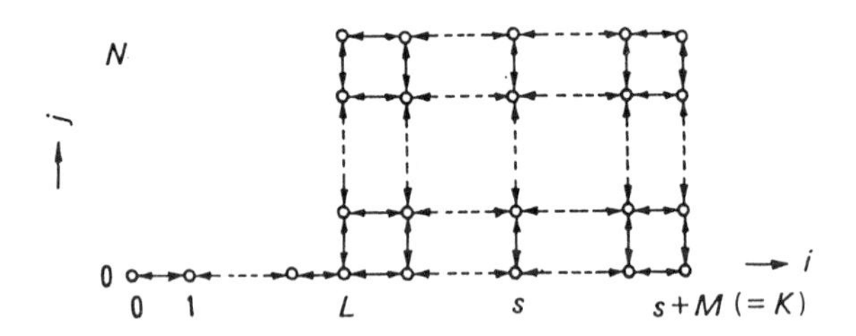

Fig.4.12 State transition diagram

Let i be the total number of busy servers and waiting priority calls, and j the number of waiting ordinary calls; and denote the system state by (i, j). The state transition mode is illustrated in Figure 4.12. Since finite waiting rooms are assumed, a steady state is always exists. Letting P_{ij} be the probability of the state (i, j), we have the steady state equation,

$$(\lambda_1 + \lambda_2 + i\mu)P_{i0}$$
$$= \delta^c_{i0}(\lambda_1 + \lambda_2)P_{i-1,0} + (i+1)\mu P_{i+1,0}, \quad 0 \le i \le L-1 \tag{4.37}$$

$$(\delta^c_{KL}\lambda_1 + \delta^c_{N0}\lambda_2 + L\mu)P_{L0}$$
$$= (\lambda_1 + \lambda_2)P_{L-1,0} + \delta^c_{KL}\min(L+1, s)\mu P_{L+1,0} + \delta^c_{N0}L\mu P_{L1} \tag{4.38}$$

$$[\delta^c_{iK}\lambda_1 + \delta^c_{N0}\lambda_2 + \min(i, s)\mu]P_{i0}$$
$$= \lambda_1 P_{i-1,0} + \delta^c_{iK}\min(i+1, s)\mu P_{i+1,0}, \quad L+1 \le i \le K \tag{4.39}$$

$$(\delta^c_{KL}\lambda_1 + \delta^c_{jN}\lambda_2 + L\mu)P_{ij}$$
$$= \lambda_2 P_{L,j-1} + \delta^c_{KL}\min(L+1, s)\mu P_{L+1,j} + \delta^c_{jN}L\mu P_{L,j+1}, \quad 1 \le j \le K \tag{4.40}$$

$$[\lambda_1 + \delta^c_{jN}\lambda_2 + \min(i, s)\mu]P_{i0}$$

$$= \lambda_2 P_{i,j-1} + \min(i+1,s)\mu P_{L+1,j} + \lambda_1 P_{i-1,j}, \quad L+1 \le i \le K-1 \tag{4.41}$$

$$[\delta^c_{jN}\lambda_2 + \min(K,s)\mu]P_{Kj} = \lambda_1 P_{K-1,j} + \lambda_2 P_{K,j-1}, \quad 1 \le j \le N. \tag{4.42}$$

where $L = s - R$, $K = s + M$, and

$$\delta^c_{ij} = \begin{cases} 0, & i = j \\ 1, & i \neq j. \end{cases}$$

Delete (4.40) to (4.42) if $N = 0$; delete (4.41) if $K \le L+1$; and delete (4.39), (4.41) and (4.42) if $K = L$.

The steady state probabilities are calculated by (4.37) to (4.41) together with the normalization condition.

4.3.3 Individual Performance Measures

The steady state probabilities are calculated recursively by the following algorithm:

[Algorithm 4.1]

(1) Set $P_{00} = C$ (for example 1).

(2) Compute (2-1) and (2-2) below, repeating from $j = 0$ to N (which is abbreviated as $0 \le j \le N$ in the sequel).

(2-1) (i) If $j = 0$, then compute

$$P_{i+1,0} = [(\lambda_1 + \lambda_2 + i\mu)P_{i0} - \delta^c_{i0}(\lambda_1 + \lambda_2)P_{i-1,0}]/[(i+1)\mu], \quad 0 \le i \le L-1.$$

(ii) If $j = 1$, then compute

$$P_{L1} = [(\delta^c_{KL}\lambda_1 + \lambda_2 + L\mu)P_{L0} - (\lambda_1 + \lambda_2)P_{L-1,0} - \delta^c_{KL}\min(L+1,s)\mu P_{l+1,0}]/(L\mu).$$

(iii) If $j \ge 2$, then compute

$$P_{Lj} = [(\delta^c_{KL}\lambda_1 + \lambda_2 + L\mu)P_{L,j-1} - \lambda_2 P_{L,j-2} - \delta^c_{KL}\min(L+1,s)\mu P_{L+1,j-1}]/(L\mu).$$

(2-2) (i) If $K = L$ skip this step, and go to (3).

(ii) If $K \neq L$, then compute A_k and B_k for $2 \le k \le K - L$ as follows:

$$A_1 = \lambda_1 + \delta_{jN}^c \lambda_2 + \min(L+1, s)\mu$$

$$B_1 = \lambda_1 + P_{Lj} + \delta_{j0}^c \lambda_2 P_{L+1,j-1}$$

$$\cdots$$

$$A_k = \delta_{k,K-L}^c \lambda_1 + \delta_{jN}^c \lambda_2 + \min(L+k, s)\mu(1 - \lambda_1/A_{k-1})$$

$$B_k = \lambda_1 B_{k-1}/A_{k-1} + \delta_{j0}^c \lambda_2 P_{L+k,j-1}.$$

(iii) Using the results in (ii) above, calculate

$$P_{ij} = [B_{i-L} + \delta_{iK}^c \min(i+1, s)\mu P_{i+1,j}]/A_{i-L}$$

for $i = K, K-1, \cdots, L+1$ (in reverse order).

(3) Calculate the total sum,

$$P = \sum_{i=0}^{L-1} P_{i0} + \sum_{j=0}^{N} P_{ij}.$$

(4) Normalize by letting

$$P_{ij}/P \to P_{ij}.$$

Using P_{ij}, the mean number L_r of waiting calls, mean waiting time W_r, and blocking probability (buffer overflow probability) B_r for the respective inputs, are given by

$$\begin{aligned} L_1 &= \sum_{i=s+1}^{K} (i-s) \sum_{j=0}^{N} P_{ij}, \quad L_2 = \sum_{j=1}^{N} j \sum_{j=L}^{K} P_{ij} \\ B_1 &= \sum_{j=0}^{N} P_{Kj}, \quad B_2 = \sum_{i=L}^{K} P_{iN} \end{aligned} \tag{4.43}$$

where $r = 1$ for the priority call and $r = 2$ for the ordinary call. From the Little formula, we have

$$W_r = \frac{L_r}{(1-B_r)\lambda_r}, \quad r = 1, 2. \tag{4.44}$$

[Example 4.7] Figure 4.13 shows a numerical example for the delay-delay system, where $h = \mu^{-1}$ is the mean service time, $a_1 = \lambda_1 h$ and $a_2 = \lambda_2 h$ the offered loads of priority and ordinary traffic, respectively, with $a = a_1 + a_2$ and $Y = a_1/a_2$. It can be seen that predetermined individual GOS's will be satisfied by the trunk reservation scheme.

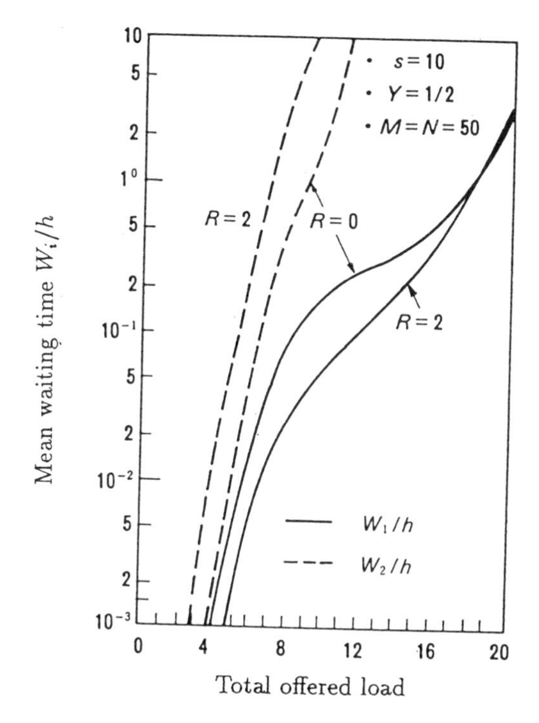

Fig.4.13 Mean waiting time of trunk reservation system (delay-delay)

4.4 Mixed Loss and Delay Systems

A system which integrates calls requiring real-time service such as voice, and deferrable calls like data, is called a *mixed loss and delay system*. Such systems appear in ISDN (integrated services digital networks) where various media shear the network resources. Typical models with simplified solutions are presented here. In the sequel, we refer to calls served on a loss basis and on a delay basis, as non-delay and delay calls (traffic), respectively.

4.4.1 $M_1+M_2/M/s(0,\infty)$

Consider the system shown in Figure 4.14(a) in which non-delay and delay calls with Poisson arrival rates λ_1 and λ_2, respectively, and common exponential service time with mean μ^{-1}, are offered to s servers. It is assumed that if all servers are busy upon call arrival, a non-delay call is lost whereas a delay call waits in an infinite waiting room for FIFO service.

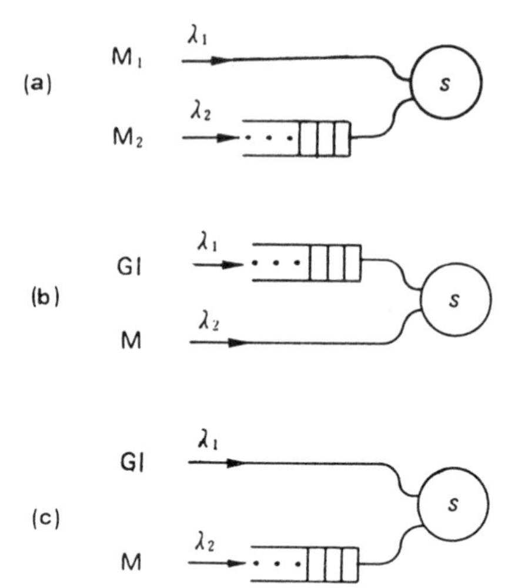

Fig.4.14 Mixed loss and delay system

It is known that a steady state exists if and only if the offered delay traffic load is $a_2 = \lambda_2/\mu < s$. Letting i be the number of busy servers, and j the number of waiting calls, we have the steady state equation,

$$(\lambda_1 + \lambda_2 + i\mu)P_{i0} = \delta^c_{0i}(\lambda_1 + \lambda_2)P_{i-1,0} + (i+1)\mu P_{i+1,0}, \quad 0 \le i \le s-1 \tag{4.45}$$

$$(\lambda_2 + s\mu)P_{s0} = (\lambda_1 + \lambda_2)P_{s-1,0} + s\mu P_{s1} \tag{4.46}$$

$$(\lambda_2 + s\mu)P_{sj} = \lambda_2 P_{s,j-1} + s\mu P_{s,j+1}, \quad j \ge 1 \tag{4.47}$$

where

$$\delta^c_{ij} = \begin{cases} 0, & i = j \\ 1, & i \ne j. \end{cases}$$

From (4.45), we have for $0 \le i \le s-1$

$$\begin{aligned} &(i+1)\mu P_{i+1,0} - (\lambda_1 + \lambda_2)P_{i0} \\ &\quad = i\mu P_{i0} - (\lambda_1 + \lambda_2)P_{i-1,0} \\ &\quad = \cdots \\ &\quad = \mu P_{10} - (\lambda_1 + \lambda_2)P_{00}. \end{aligned} \tag{4.48}$$

From this we obtain the recurrence formula,

$$P_{i+1,0} = \frac{\lambda_1 + \lambda_2}{(i+1)\mu} P_{i0}, \quad 0 \le i \le s-1. \tag{4.49}$$

Successively applying (4.49), and setting $a = (\lambda_1 + \lambda_2)/\mu$, we get

$$P_{i0} = \frac{a^i}{i!} P_{00}, \quad 1 \le i \le s. \tag{4.50}$$

To obtain P_{sj}, introduce the *marginal generating function*,

$$g(z,s) = \sum_{j=0}^{\infty} z^j P_{sj}. \tag{4.51}$$

From (4.46) and (4.47), we have

$$\left[a_2(1-z) + s\left(1 - \frac{1}{z}\right)\right] g(z,s) = a_2 P_{s-1,0} - \frac{sP_{s0}}{z} \tag{4.52}$$

where $a_2 = \lambda_2/\mu$ is the offered load of delay traffic. Using (4.50) in (4.52) yields

$$g(z,s) = \frac{1}{s - a_2 z} \frac{a^s}{(s-1)!} P_{00}. \tag{4.53}$$

From the normalization condition,

$$\sum_{i=0}^{s-1} P_{i0} + g(1,s) = 1 \tag{4.53a}$$

we have

$$P_{00} = \left[\sum_{i=0}^{s-1} \frac{a^i}{i!} + \frac{a^s}{(s-1)!} \frac{1}{s-a_2}\right]^{-1} = \frac{(s-a_2)E_s(a)}{s - a_2 + a_2 E_s(a)} \frac{s!}{a^s} \tag{4.54}$$

where $E_s(a)$ is the Erlang B formula defined in (2.26).

Expanding (4.53) in a power series of z, from the coefficient of z^j, we have

$$P_{sj} = \frac{a^s}{s!} \left(\frac{a_2}{s}\right)^j P_{00}. \tag{4.55}$$

The blocking probability of the non-delay traffic is given by

$$B = g(1,s) = \frac{1}{s-a_2} \frac{a^s}{(s-1)!} P_{00} = \frac{sE_s(a)}{s - a_2 + a_2 E_s(a)}. \tag{4.56}$$

From the PASTA, the waiting probability of the delay traffic is the same as B.

Next, we shall derive the waiting time of the delay traffic. Let $V(t)$ be the distribution function of the residual time until a server becomes idle when all servers are busy, and $v^*(\theta)$ be its LST. Then, since the service time of all calls is identically exponentially distributed, we have

$$v^*(\theta) = \frac{n\mu}{\theta + n\mu}. \tag{4.56a}$$

Letting $W(t)$ be the waiting time distribution and $w^*(\theta)$ its LST, for FIFO discipline, we have

$$\begin{aligned} w^*(\theta) &= \sum_{i=0}^{s-1} P_{i0} + \sum_{j=0}^{\infty} P_{sj}[v^*(\theta)]^{j+1} \\ &= 1 - B + B\left(1 - \frac{a_2}{s}\right)\sum_{j=0}^{\infty}\left(\frac{a_2}{s}\right)^j \left(\frac{s\mu}{\theta + s\mu}\right)^{j+1} \\ &= 1 - B\left[1 - \frac{s - a_2}{\theta + (s - a_2)\mu}\right]. \end{aligned} \tag{4.56b}$$

Inverting (4.56b), the waiting time distribution function is given by

$$W(t) = 1 - Be^{-(s-a_2)\mu t}. \tag{4.57}$$

The mean waiting time is given by

$$W = \frac{B}{s - a_2} h \tag{4.58}$$

where $h = \mu^{-1}$ is the mean service time.

If the service times are different for non-delay and delay calls, the analyses are made in [19] and [20], but such simplified solutions as those above are not obtained.

4.4.2 GI+M/M/$s(\infty, 0)$

When delay calls arrive in a renewal process (GI) as shown in Figure 4.11(b), an approximate solution using the *renewal approximation* is proposed in [21]. The outline of the result is summarized below.

Let $A_1(t)$ and $A_2(t)$ be the interarrival time distribution functions of 2 independent renewal inputs with respective arrival rates λ_1 and λ_2. Although the mixed input process of these is not renewal in general, if regarding it as renewal, its interarrival time distribution function $A(t)$ is given by [22]

$$A^c(t) = \frac{\lambda_1 \lambda_2}{\lambda}\left[A_1(t)\int_t^{\infty} A_2^c(x)dx + A_2(t)\int_t^{\infty} A_1^c(x)dx\right] \tag{4.59}$$

where $\lambda = \lambda_1 + \lambda_2$ is the total arrival rate. If $A_2(t)$ is an exponential distribution, *i.e.* Poisson (M) input, from (4.59) we obtain the LST of $A(t)$,

$$a^*(\theta) = \frac{\lambda_1}{\lambda}\phi_1(\theta) + \frac{\lambda_2}{\lambda}\phi_2(\theta) \tag{4.59a}$$

where

$$\phi_1(\theta) = \frac{\lambda_2 + \theta a_1^*(\theta + \lambda_2)}{\theta + \lambda_2}$$
$$\phi_2(\theta) = \frac{\lambda_2(\theta + \lambda_2) + \theta\lambda_1[1 - a_1^*(\theta + \lambda_2)]}{(\theta + \lambda_2)^2} \tag{4.59b}$$

with $a_1^*(\theta)$ being the LST of the interarrival time distribution $A_1(t)$ of the GI input.

Analysing GI/M/$s(\infty, 0)$ with the mixed renewal input, and using the PASTA, the mean waiting time W_1 for the delay GI input, and the blocking probability B_2 for the non-delay Poisson (M) input are approximated as

$$W_1 = \frac{\lambda M(0) - \lambda_2 B_2}{\lambda_1 s(1-\omega)} h$$
$$B_2 = M(0)\frac{\lambda_1\phi_1(s\mu) + \omega\lambda_2\phi_2(s\mu)}{(s + \lambda_2 h)\omega a^*(s\mu)} \tag{4.60}$$

where $h = \mu^{-1}$ is the mean service time.

The waiting probability $M(0)$ is given by

$$M(0) = \left[1 + (1-\omega)\sum_{r=1}^{s}\binom{s}{r}\frac{s[1 - \hat{a}^*(r\mu)] - r}{[s(1-\omega) - r][1 - \hat{a}^*(r\mu)]}\prod_{i=1}^{r}\frac{1}{\phi(i\mu)}\right]^{-1} \tag{4.61}$$

where $\phi(\theta) = a^*(\theta)/[1 - a^*(\theta)]$, and $\hat{a}^*(\theta)$ is the LST of the *modified interarrival time* due to lost calls, which is given by

$$\hat{a}^*(\theta) = \frac{\lambda_1\phi_1(\theta)}{\lambda - \lambda_2\phi_2(\theta)}. \tag{4.62}$$

The generalized occupancy ω is given similarly to (3.80) as a root $(0 < \omega < 1)$ of the equation,

$$\omega = \hat{a}^*([1-\omega]s\mu). \tag{4.62a}$$

4.4.3 GI+M/M/$s(0, \infty)$

For the mixed system shown in Figure 4.11(c), the blocking probability B_1 for non-delay GI input, and the mean waiting time W_2 for delay M input are similarly approximated by [21]

$$B_1 = M(0)\frac{\phi_1(s\mu)}{a^*(s\mu)}$$
$$W_2 = M(0)\frac{\phi_2(s\mu)}{(s - \lambda_2 h)a^*(s\mu)} h \tag{4.63}$$

with the same notation as before except for the waiting probability given by

$$M(0) = \left[1 + \left(1 - \frac{\lambda_2 h}{s}\right) \sum_{r=1}^{s} \binom{s}{r} \prod_{i=1}^{r} \frac{1}{\phi(i\mu)}\right]^{-1}. \tag{4.64}$$

If the GI input becomes Poissonian, B and W in (4.60) or (4.63) coincide with the exact solutions in (4.56) and (4.58), respectively.

[Example 4.8] With the GI input of H_2 (hyper-exponential distribution), we have

$$a_1^*(\theta) = \frac{kr_1}{\theta + r_1} + \frac{(1-k)r_2}{\theta + r_2}. \tag{4.64a}$$

With the symmetric H_2, the arrival rate $\lambda_1 = 2$/sec and the SCV ${C_a}^2 = 2$ of the interarrival time, from (3.102) we have

$$k = 0.7887, \quad r_1 = 3.1548/\text{sec}, \quad r_2 = 0.8452/\text{sec}.$$

For the H_2+M/M/$s(0,\infty)$ with the non-delay H_2 input and delay M input of $\lambda_2 = 2$/sec, letting $h = 1$ sec and $s = 5$, from (4.63) we have

$$B_1 = 0.3051, \quad W_2 = 0.0970\,\text{sec}.$$

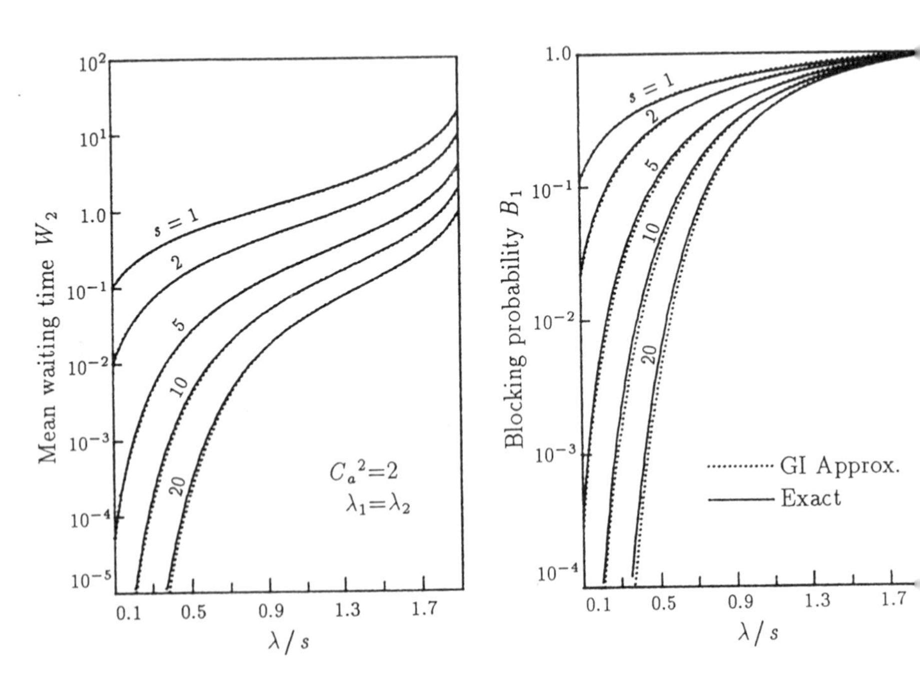

Fig.4.15 Calculated example for H_2+M/M/$s(0,\infty)$

In Figure 4.15, a calculated example is compared with the exact solution, which will be obtained by the matrix analytic method in Section 6.2. (See Example 6.4.) The approximation presents a fairly good accuracy in a relatively simple calculation.

4.5 Multi-Queue Models

4.5.1 Modeling of Multi-Queue

A system in which a number of queues are served according to a certain discipline, is called a *multi-queue model*. Figure 4.16 shows a cyclic-type multi-queue model which appears in the polling system in data communication, where a controller examines the call request of each station (queue), to process it, if any, and then moves to the next station sequentially. Such a model is also found in the Token Ring LAN (local area network), which passes the token along with nodes (queues) connected in a ring-type transmission path, and only the node that gets the free token, is allowed to transmit data [7] [23]. (See Subsection 6.5.1.)

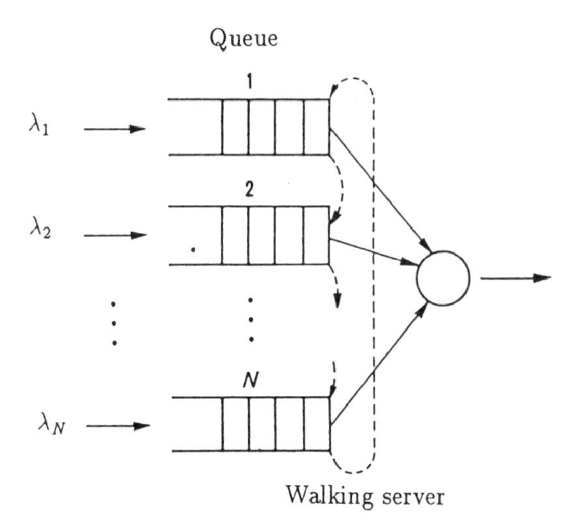

Fig.4.16 Multi-queue model (cyclic-type)

The multi-queue model is classified by the disciplines serving the queues, as shown in Figure 4.17.

(a) *Exhaustive model*: Server moves to the next queue, after having served all calls (requests) in a queue including those arriving in the service period.

(b) *Gated model*: Server moves to the next queue, after having served calls waiting in a queue, which arrived before the service commenced.

(c) *Limited model*: Server moves to the next queue, after having served a limited number of calls waiting in a queue plus calls arriving in the service time. If the first k calls are served, it is called an k-limited model.

Furthermore, according to the discipline of server moving around the queues, the model is classified as *cyclic-type*, *stochastic type*, or *priority type*.

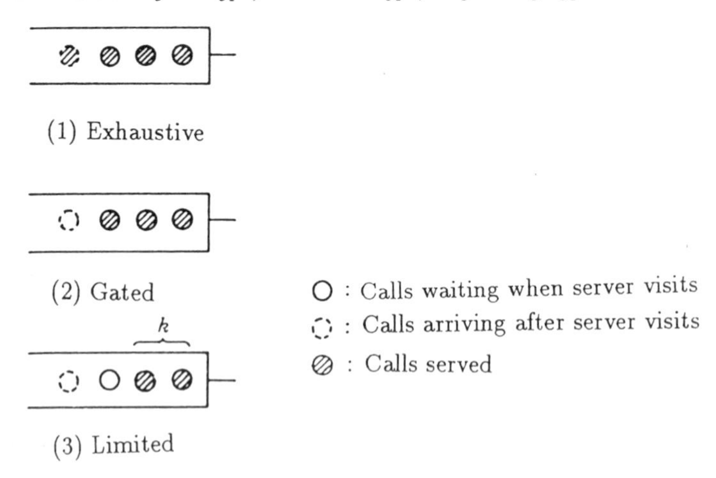

Fig.4.17 Multi-queue serving disciplines

The time taken by the server to move between queues, is called the *walking time* or *moving time*. If the traffic characteristics (arrival process, service time, etc.) and the walking time for each queue are identical, then the model is called *symmetric*, otherwise *asymmetric*.

This section deals with cyclic-type models, and the following notation is used:

λ_i : Arrival rate to queue i.
h_i : Mean service time of queue i.
$h_i^{(2)}$: Second moment of service time of queue i.
u_i : Mean walking time from queue i to the next queue.
$u_i^{(2)}$: Second moment of walking time from queue i to the next queue.
σ_{ui}^2 : Variance of walking time from queue i to the next queue.

The *cycle time* is defined as the time interval from the instant at which the server begins to serve a queue (station) until the server visits the same queue again. Letting a_i be the probability that a call exists in queue i when the server visits the queue, and the c_i be the mean cycle time of the queue, then we have

$$a_i = \lambda_i c_i. \tag{4.65}$$

Letting N be the number of queues, and setting

$$c_0 = \sum_{i=1}^{N} u_i \tag{4.65a}$$

which is the total net walking time, it follows that

$$c_i = c_0 + \sum_{i=1}^{N} a_i h_i. \tag{4.66}$$

Using (4.65) in (4.66), and solving for c_i, we have

$$c_i = \frac{c_0}{1-\rho_0} \equiv c \tag{4.67}$$

where

$$\rho_0 = \sum_{i=1}^{N} \lambda_i h_i$$

is the server utilization. Since c_0 and ρ_0 are independent of i, so is the mean cycle time, which is denoted by c.

4.5.2 Exhaustive Model

The model with Poisson arrivals and general service time for all queues, is called the *M/G/1-type model.* The time interval from when the server leaves a queue until visits the same queue again, is called the *inter-visit time.* In an M/G/1-type exhaustive model, the mean waiting time for queue i is given by

$$W_{Ei} = \frac{v_i^{(2)}}{2v_i} + \frac{\lambda_i h_i^{(2)}}{2(1-\rho_i)} \tag{4.68}$$

where $\rho_i = \lambda_i h_i$ is the offered load, and v_i and $v_i^{(2)}$ are, respectively, the mean and second moment of the inter-visit time for queue i. Equation (4.68) may be interpreted as follows: The mean waiting time is the sum of the mean residual inter-visit time (mean time until server visits the queue) and the mean waiting time of the M/G/1 system in the queue. Letting $c_i^{(2)}$ be the second moment of the cycle time of queue i, (4.68) can be rewritten as [45]

$$W_{Ei} = (1-\rho_i)\frac{c_i^{(2)}}{2c}. \tag{4.69}$$

For the symmetric model, omitting the suffix i, we have

$$v = \frac{(1-\rho)c_0}{1-\rho_0} \tag{4.70}$$

$$v^{(2)} = \frac{N\sigma_u^2(1-\rho)}{1-\rho_0} + \frac{N(N-1)\lambda u h^{(2)}}{(1-\rho_0)^2} + v^2 \tag{4.71}$$

$$c^{(2)} = N\frac{\sigma_u^2 + \lambda h^{(2)} c}{(1-\rho)(1-\rho_0)} + c^2. \tag{4.72}$$

where $\rho = \lambda h$ is the offered load per queue (station). Substituting (4.70) and (4.71) into (4.68), or (4.72) into (4.69), the mean waiting time is given by

$$W_E = \frac{\sigma_u^{(2)}}{2u} + \frac{(1-\rho)c_0 + N\lambda h^{(2)}}{2(1-\rho_0)}. \tag{4.73}$$

The exact solution has been also presented for the asymmetric model, but it requires an iterative numerical solution of an equation system [45]. A simple approximate solution is given in [46]. This will be described in Subsection 6.5.2.

4.5.3 Gated Model

The mean waiting time for the M/G/1-type gated model is given by [45]

$$W_{Gi} = (1+\rho_i)\frac{c_i^{(2)}}{2c} \tag{4.74}$$

where the notation is the same as before. For the symmetric system, omitting the suffix i, we have

$$c^{(2)} = N\frac{\sigma_u^2 + \lambda h^{(2)} c}{(1+\rho)(1-\rho_0)} + c^2. \tag{4.75}$$

Substituting this into (4.74) we obtain the mean waiting time,

$$W_G = \frac{\sigma_u^{(2)}}{2u} + \frac{(1+\rho)c_0 + N\lambda h^{(2)}}{2(1-\rho_0)}. \tag{4.76}$$

It should be noted that only difference from (4.73) for the exhaustive model is the sign in the numerator of the second term. The exact solution for the asymmetric model has also been presented in [45]. For practical application, the approximate formula in Subsection 6.5.3 is also useful.

4.5.4 Limited Model

The analysis for the limited model is more involved. Here, we shall present the result for the symmetric M/G/1-type 1-limited model [24].

Choose an arbitrary call as the test call, and let Q_d be the mean number of calls arriving in the sojourn time of the test call. Then, Q_d is equal to the mean number of calls existing in the system just after the test call departure. Thus, letting W_L be the mean waiting time, from the Little formula, we have

$$Q_d = \lambda(W_L + h). \tag{4.77}$$

By introducing the joint probability generating function for individual queues, we can derive

$$Q_d = \lambda \left[h + \frac{\sigma_u^2}{2u} + \frac{N}{2} \frac{\lambda h^{(2)} + u(1+\rho) + \lambda \sigma_u^2}{1 - N\lambda(u+h)} \right]. \tag{4.78}$$

Using (4.78) in (4.77), and solving for W_L, we obtain

$$W_L = \frac{\sigma_u^2}{2u} + \frac{N}{2} \frac{\lambda h^{(2)} + u(1+\rho) + \lambda \sigma_u^2}{1 - N\lambda(u+h)}. \tag{4.79}$$

Exercises

[1] Consider the batch arrival waiting system $M^{[X]}/D/1$ with arrival rate 0.3/sec, mean batch size 2 and fixed service time 1 sec.

(1) Calculate the mean waiting times for geometric and fixed batch sizes.

(2) Compare the mean waiting time with that of non-batch arrival M/D/1 with the same offered load.

[2] Classify computer jobs in 5 non-preemptive priority classes. Assume for each class, Poisson arrival rate of 10 jobs/sec, and the processing time of mean 10 ms and second moment 120 ms^2. Then, calculate

(1) The mean waiting times for each class.

(2) The mean waiting time for all the jobs.

[3] With the parameters in Example 4.8, calculate the blocking probability and mean waiting time:

(1) Exact value for $M_1+M_2/M/s(0,\infty)$.

(2) Approximate value for $H_2+M/M/s(\infty,0)$.

[4] In the symmetric cyclic-type queue with 10 stations each having Poisson arrival rate 0.05/sec, exponential service time with mean 1 sec and fixed walking time 0.5 sec, calculate

(1) The mean waiting time for the exhaustive model.

(2) The mean waiting time for the 1-limited model.

Chapter 5

ALTERNATIVE ROUTING SYSTEMS

In the alternative routing system, if all servers are busy in a certain route, blocked traffic is carried by an alternative route. The alternative routing system is a limited availability system, and various practical approximations have been studied, as well as the exact solution by numerical analysis. The typical methods are described in this chapter.

5.1 Overflow System

5.1.1 Overflow Traffic Model

Consider the overflow traffic model shown in Figure 5.1. If Poissonian arrivals at rate λ find the primary group of all the s servers busy, blocked traffic overflows to the secondary group with infinite servers.

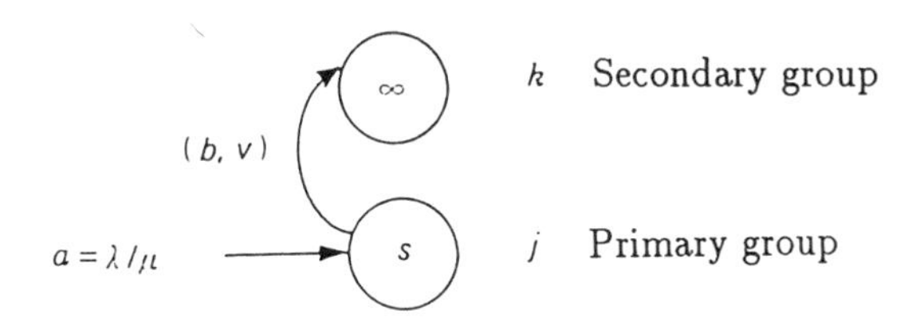

Fig.5.1 Overflow traffic model

Assuming the service time is exponentially distributed with mean μ^{-1} in both groups, we have the state transition diagram as shown in Figure 5.2. Letting P_{jk} be the steady state probability that j and k calls exist in the primary and secondary groups, respectively, from the rate-out=rate-in we obtain the steady state equation,

$$[\lambda+(j+k)\mu]P_{jk}=\lambda P_{j-1,k}+(j+1)\mu P_{j+1,k}+(k+1)\mu P_{j,k+1}$$
$$j=0,1,\cdots,s-1;\quad k=0,1,\cdots;\quad P_{-1,k}=0 \tag{5.1}$$

$$[\lambda+(s+k)\mu]P_{sk}=\lambda P_{s-1,k}+\lambda P_{s,k-1}+(k+1)\mu P_{s,k+1},$$
$$j=s;\quad k=0,1,\cdots;\quad P_{s,-1}=0. \tag{5.2}$$

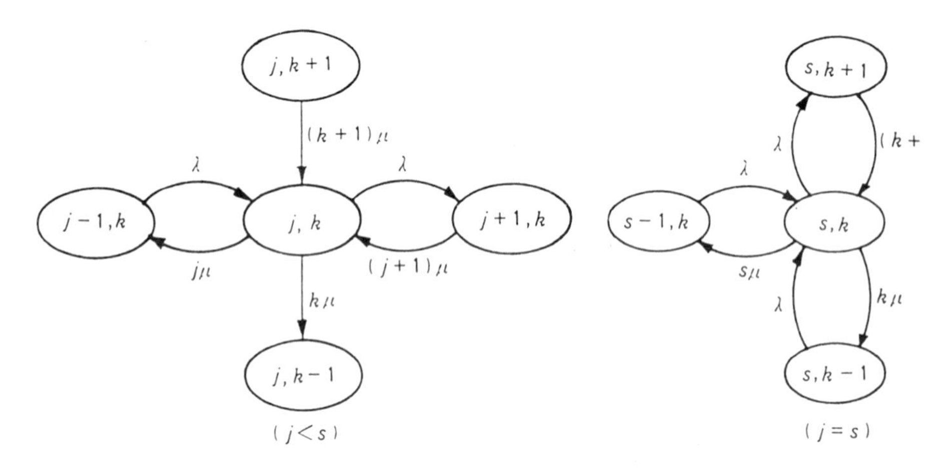

Fig.5.2 State transition diagram for overflow model

To solve the state equation, we introduce the auxiliary function Q_{jk} which satisfies (5.1) for $j=0,1,\cdots,$ as well as

$$Q_{jk}=P_{jk},\quad j=0,1,\cdots,s;\quad k=0,1,\cdots \tag{5.3}$$

and its probability generating function,

$$g(x,y)=\sum_{j=0}^{\infty}\sum_{k=0}^{\infty}x^j y^k Q_{jk}.$$

Then, from (5.1) we have the partial differential equation,

$$(1-x)\frac{\partial}{\partial x}g(x,y)+(1-y)\frac{\partial}{\partial y}g(x,y)=a(1-x)g(x,y) \tag{5.4}$$

where $a=\lambda/\mu$ is the offered load. The solution of (5.4) can be given in the form,

$$g(x,y) = \sum_{r=0}^{\infty} c_r (1-y)^r \sum_{k=0}^{\infty} \phi_r(j) x^j \tag{5.5}$$

where $\phi_r(j)$ is defined by

$$\begin{aligned} \phi_0(j) &= e^{-a}\frac{a^j}{j!} \\ \phi_{r+1}(j) &= \sum_{k=0}^{j} \phi_r(k), \quad r = 1, 2, \cdots. \end{aligned} \tag{5.6}$$

The coefficient c_r in (5.5) is determined so as to satisfy the boundary condition (5.2) and the normalization condition,

$$\sum_{j=0}^{\infty}\sum_{k=0}^{\infty} Q_{jk} = 1.$$

Thus, it follows that [4, p.66]

$$c_r = (-1)^r \frac{a^r}{r!} \frac{\phi_0(s)}{\phi_{r+1}(s)\phi_r(s)}. \tag{5.7}$$

5.1.2 Moment and LST of Overflow Process

Since the secondary group is infinite, the calls existing in the group represent those in the overflow process, which is referred to as *overflow calls* in the sequel. Hence, the nth factorial moment M_n of the number of overflow calls is given by

$$M_n = \lim_{x\to 0, y\to 1} \frac{1}{j!} \sum_{j=0}^{s} \frac{\partial^j \partial^n}{\partial x^j \partial y^n} g(x,y) = a^n \frac{\phi_0(s)}{\phi_n(s)} \tag{5.8}$$

from which we have the *Heffes formula*,

$$\begin{aligned} M_{n+1} &= a\delta_n M_n, \quad M_0 = 1 \\ \delta_n &= \frac{n}{s+n-a+a\delta_{n-1}}, \quad \delta_0 = E_s(a) \end{aligned} \tag{5.9}$$

where $E_s(a)$ represents the Erlang B formula defined in (2.26).

It is known that the inter-overflow time is renewal, and its LST is given by the *Descloux formula* [26],

$$\begin{aligned} \alpha_0(\theta) &= \frac{\lambda}{\theta+\lambda} \\ \alpha_s(\theta) &= \frac{\lambda}{\theta+\lambda+s\mu[1-\alpha_{s-1}(\theta)]}. \end{aligned} \tag{5.9a}$$

In particular, with $s = 1$ the LST is given by

$$
\begin{aligned}
a_1^*(\theta) &= \frac{k\lambda_1}{\theta+\lambda_1} + \frac{(1-k)\lambda_2}{\theta+\lambda_2} \\
\left.\begin{matrix}\lambda_1\\ \lambda_2\end{matrix}\right\} &= \frac{2\lambda+\mu\pm\sqrt{\mu(4\lambda+\mu)}}{2}, \quad k = \frac{\lambda(\lambda-\lambda_1)}{\lambda_1(\lambda_2-\lambda_1)}.
\end{aligned}
\tag{5.9b}
$$

Thus, the inter-overflow time from a single server is distributed with H_2 (2nd order hyper-exponential distribution). It is known that the inter-overflow time from s server is distributed with H_{s+1}.

It is worth noting here that the overflow process from the mixed delay and non-delay system in Figure 4.11(a) is not renewal, and the nth factorial moment of the overflow calls are given by [27]

$$
M_n = \frac{a_1 n(s-a_2)\phi_0(s)}{s\phi_1(s) - a_2\phi_1(s-1)} \left[\frac{\phi_1(s)}{\phi_n(s)} + \frac{a_2}{s-a_2}\right] \tag{5.9c}
$$

where a_1 and a_2 are the offered load of the non-delay and delay inputs, respectively. With $a_1 = a$ and $a_2 = 0$, (5.9c) reduces to (5.8).

5.1.3 Mean and Variance of Overflow Calls

From (5.9), the mean $b = M_1$ and variance $v = M_2 + M_1 - M_1^2$ of the overflow calls are given by

$$
\begin{aligned}
b &= aE_s(a) \equiv b(s,a) \\
v &= b\left(1 + \frac{a}{s+1-a+b}\right) \equiv v(s,a)
\end{aligned}
\tag{5.10}
$$

which is called the *Wilkinson formula*.

As will be seen later, in practical design, it is necessary to find a and s for given b and v. This may be done, for example, by inverse calculation of (5.10) by iteration, for example. The following *Rapp formula* is useful for facilitating the calculation in a closed form [29]:

$$
\begin{aligned}
a &\doteq v + 3z(z-1) \\
s &= a\frac{b+z}{b+z-1} - b - 1
\end{aligned}
\tag{5.11}
$$

where $z = v/b$ is called the *index of variance*.

[Example 5.1] Table 5.1 shows numerical examples of (5.10) and (5.11). It can be seen that the Rapp formula provides a fairly good approximation in a simple calculation.

Table 5.1 Accuracy of Rapp formula

b	$z = 1.5$		$z = 2.0$	
	a	s	a	s
1	3.750 3.665	4.250 3.608	8.000 7.605	10.000 9.408
5	9.750 9.463	5.523 5.184	16.000 15.356	12.667 11.915
20	32.250 32.120	12.823 12.686	46.000 45.642	27.191 26.815
50	77.250 77.187	27.780 27.715	106.000 105.812	57.078 56.887

Upper: Rapp formula Lower: Exact

5.2 Approximate Methods for Overflow Systems

5.2.1 Equivalent Random Theory

Consider the alternative routing system shown in Figure 5.3, where overflow traffic from the load a to the *high-usage route* A→C is carried through the *alternative route* A→B→C, if all the trunks (servers) in route A→C are busy. It is assumed that the load a consists of Poisson arrivals with exponential service time. In the following, Poisson arrival traffic is referred to as *random traffic*, otherwise *non-random traffic*.

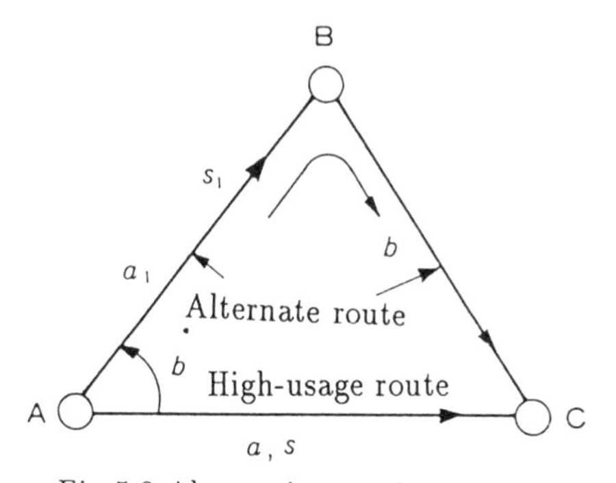

Fig.5.3 Alternative routing system

The system in consideration is modeled as in Figure 5.4(a). Letting s be the number of trunks in the high usage route, the mean b and variance v of the overflow calls to route A→B are given by

$$b = b(s, a), \quad v = v(s, a) \tag{5.12}$$

where $b(\cdot, \cdot)$ and $v(\cdot, \cdot)$ are defined in (5.10). The mean and variance are identical for random traffic (See Table C.1 in Appendix C.), but this is not the case for non-random overflow traffic. Thus, we denote a non-random traffic by (b, v).

Let a_1 be the background random traffic load inherently offered to route A→B with the same exponential service time, and denote it by (a_1, a_1). Then, assuming that (b, v) and (a_1, a_1) are independent, the superposed traffic is denoted by (b^*, v^*), where

$$b^* = a_1 + b, \quad v^* = a_1 + v. \tag{5.13}$$

From (5.12), we can uniquely determine the *equivalent random traffic* load a^* and the number s^* of *fictitious primary trunks*, which produce the non-random traffic (b^*, v^*) as the overflow traffic. They may be approximated by the Rapp formula in (5.11).

Letting s_1 be the number of trunks in route A→B, the system is approximated as shown in Figure 5.4(b), which is equivalent to a loss system with random traffic a^* offered to $(s^* + s_1)$ trunks. Therefore, the mean overflow traffic b_1 from s_1 trunks, is given by $b_1 = b(s^* + s_1, a^*)$. Hence, the blocking probability B for the overflow route, is given by

$$B = \frac{b_1}{b^*} = \frac{E_{s_1+s^*}(a^*)}{E_{s^*}(a^*)}. \tag{5.14}$$

This is called the *equivalent random theory* (ERT), and has been widely applied in the design of alternative routing systems [28].

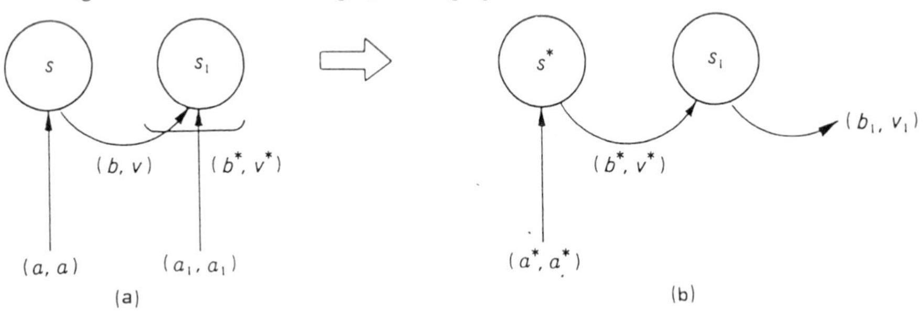

Fig.5.4 Principle of equivalent random theory

[Example 5.2] Let us calculate the blocking probability B in route A→B shown in Figure 5.3 with $a = 5\,\text{erl}$, $s = 5$; $a_1 = 10\,\text{erl}$ and $s_1 = 20$.

From (5.12) we have $b = 1.4243$ and $v = 2.3332$, and hence $b^* = 11.4243$ and $v^* = 12.3332$ (omitting the unit erl for simplicity). Solving (5.12) numerically, we have $a^* = 12.57$, $s^* = 1.24$ and $s^* + s_1 = 21.24$. Applying the Erlang B formula for a real number of servers (See Appendix B.2.), we have

$$E_{1.24}(a^*) = 0.9089, \quad E_{21.24}(a^*) = 0.0073, \quad B = 0.0081.$$

f the Rapp formula in (5.11) is used, we obtain $a^* = 12.59$ and $s^* = 1.26$ which give he same result. Since the exact value is $B = 0.0085$ (See Table 5.5 in Section 5.4.), ıe approximation presents fairly good accuracy for practical application using a mple calculation.

The ERT gives the average blocking probability in the overflow route, but it can ›t provides the individual blocking probabilities for traffic loads a and a_1. Let us troduce the interrupted Poisson process below, for an approximate evaluation of ʟne individual performances, .

5.2.2 Interrupted Poisson Process

The overflow process is often approximated by the *interrupted Poisson process* (IPP). The IPP is shown in Figure 5.5, in which the Poisson input at rate λ is interrupted by a *random switch* with exponential on and off intervals with means γ^{-1} and ω^{-1}, respectively. The interarrival time of the IPP follows the H_2 (2nd order hyper-exponential distribution) with the LST [30],

$$a^*(\theta) = \frac{kr_1}{\theta + r_1} + \frac{(1-k)r_2}{\theta + r_2} \tag{5.15}$$

where

$$\left.\begin{matrix} r_1 \\ r_2 \end{matrix}\right\} = \frac{\lambda + \gamma + \omega \pm \sqrt{(\lambda + \gamma + \omega)^2 - 4\lambda\omega}}{2}, \quad k = \frac{\lambda - r_2}{r_1 - r_2}. \tag{5.16}$$

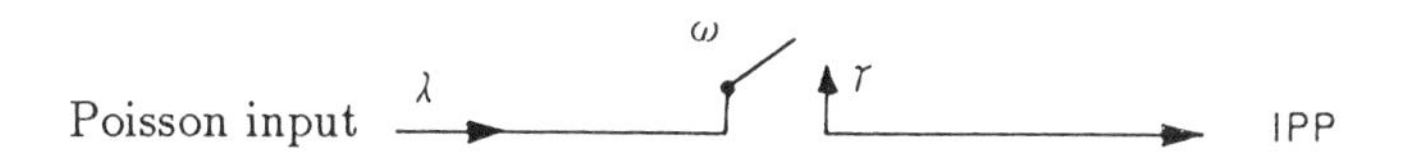

Fig.5.5 Interrupted Poisson process

As stated in Subsection 5.1.2, the inter-overflow time from s servers is distributed with H_{s+1}. In the IPP approximation, the inter-overflow time distribution is approximated by IPP with H_2 having the first 2 or 3 moments matching. Here, we shall show the 2-moment match method.

Let $z = v/b$ be the index of variance of the non-random traffic (b, v). Matching the mean and variance of the number of existing calls, and setting λ equal to the equivalent random traffic approximated by the Rapp formula (5.11), the IPP parameters are given by

$$\lambda = bz + 3z(z-1), \quad \gamma = \omega\left(\frac{\lambda}{b} - 1\right), \quad \omega = \frac{b}{\lambda}\left(\frac{\lambda - b}{z - 1} - 1\right) \tag{5.17}$$

where the unit of time is the mean service time.

Using (5.16) and (5.17) in (5.15), and setting

$$\phi(\theta) = \frac{a^*(\theta)}{1 - a^*(\theta)} \tag{5.17a}$$

we have

$$\phi(\theta) = \frac{b}{\theta}[1 + (z-1)f(\theta, b, z)] \tag{5.18}$$

where

$$f(\theta, b, z) \equiv \frac{\theta}{b} \frac{b + 3z}{\theta - 1 + b + 3z}. \tag{5.19}$$

5.2.3 GI Approximation

The superposed process of the overflow traffic and the background random traffic in the overflow route, is non-renewal. However, if it is regarded as renewal, the alternative route in Figure 5.3 is modeled by GI/M/s_1(0). Hence, from (3.73), the blocking (call congestion) probability is given by

$$B = \left[1 + \sum_{r=1}^{s_1} \binom{s_1}{r} \frac{1}{\prod_{i=1}^{r} \phi(i\mu)}\right]^{-1} \tag{5.20}$$

where $\phi(\cdot)$ is given by (5.18) with $b = b^*$, $v = v^*$ and $z = v^*/b^*$.

Denoting the blocking probabilities for the overflow traffic (b) and the background random traffic (a_1) in the alternative route by B_0' and B_1, respectively, we have the *load conservation law*

$$b^* B = bB_0' + a_1 B_1. \tag{5.21}$$

From the PASTA, B_1 is equal to the time congestion probability, and from (3.74) we have the relation,

$$B = \frac{s_1}{b^*}\phi(s_1\mu)B_1. \tag{5.22}$$

Using (5.22) in (5.21), we have the individual blocking probabilities,

$$\begin{aligned} B_0' &= [1 + (z_0 - 1)f(s_1, b^*, z^*)]B_1 \\ B_1 &= [1 + (z^* - 1)f(s_1, b^*, z^*)]^{-1}B \end{aligned} \tag{5.23}$$

where $f(\cdot,\cdot,\cdot)$ is given by (5.19) with $z_0 = v/b$ and $z^* = v^*/b^*$. The overall blocking probability B_0 for the load a offered to route A→C is given by

$$B_0 = \frac{b}{a_0} B_0'. \tag{5.24}$$

This method is called *GI approximation*, because the non-renewal superposed process is approximated by renewal (GI) process [31].

[Example 5.3] Table 5.2 shows numerical examples of ERT and GI approximation compared with the exact solution by numerical analysis which will be described in Section 5.4. The GI approximation gives the individual blocking probabilities with fairly good accuracy in a simple calculation.

Table 5.2 Accuracy of the approximate methods

a	s	a_1	s_1	B_0	B_1	B
5	5	10	20	- 0.0113 0.0132	- 0.0076 0.0079	0.0081 0.0081 0.0085
10	7	12	25	- 0.0078 0.0082	- 0.0129 0.0131	0.0144 0.0145 0.0149
15	10	15	30	- 0.0292 0.0294	- 0.0191 0.0193	0.0219 0.0220 0.0222

Upper: ERT Middle: GI Approx. Lower: Exact

5.3 Optimum Design of Alternative Routing

5.3.1 Conventional Method

In an alternative routing system, varying the number of high-usage trunks with a given GOS, the system cost changes as shown in Figure 5.6(a). Let us consider the optimum design to minimize the system cost. First, we shall describe the conventional method used in telephone networks.

Let c_0, c_1 and c_2 be the costs per trunk in the high-usage route A→C and alternative routes, A→B, and B→C respectively, as shown in Figure 5.6(b), which are estimated including the cost of switching facilities. Then, denoting the marginal costs per traffic load (erl) with respect to an additional trunk in alternative and high-usage routes by c_A and c_H, respectively, we have the relation,

$$\begin{aligned} c_A &= \frac{c_1 + c_2}{\text{ATC of alternative route}} \\ c_H &= \frac{c_0}{\text{LTC of high usage route}} \end{aligned} \tag{5.25}$$

where ATC is the additional trunk capacity, and LTC the last trunk capacity defined in Subsection 2.1.5. The provision of high-usage route is justified, if $c_H \leq c_A$.

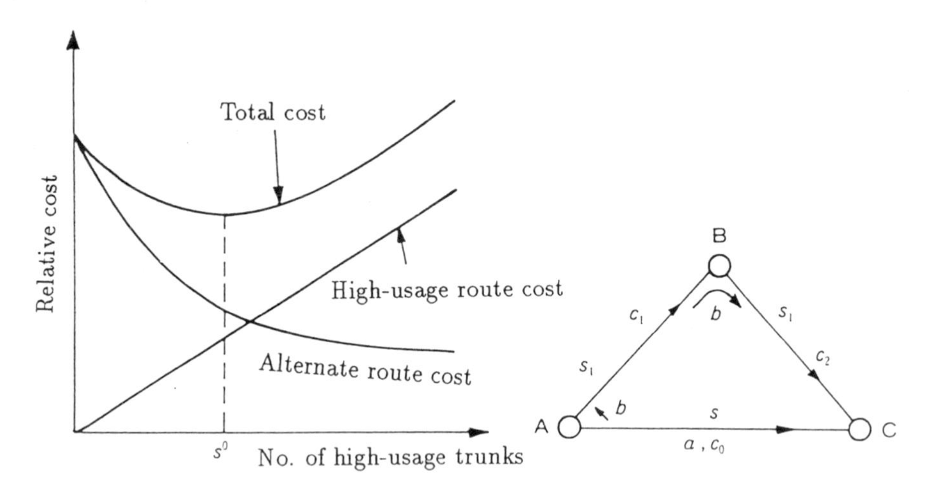

Fig.5.6 Cost of alternative routing system

Define *cost ratio* k of the alternative to high-usage route by

$$k \equiv \frac{c_1 + c_2}{c_0}. \tag{5.26}$$

Then, the condition to minimize the system cost is given by

$$\frac{\text{ATC}}{\text{LTC}} = k. \tag{5.27}$$

Letting a and s be the offered load and the number of trunks in the high-usage route, the LTC is given by

$$\text{LTC} = a[E_{s-1}(a) - E_s(a)]. \tag{5.28}$$

The ATC is assumed constant, as shown in the following example:

[Example 5.4] Let us find the optimum number of high-usage trunks for minimizing the system cost, with $k = 1.5$ and $a = 5\,\text{erl}$.

In practice ATC $= 0.83\,\text{erl}$ is used, and hence from (5.27) we have the relation,

$$\text{LTC} \geq \frac{0.83}{1.5} = 0.55\,\text{erl}. \tag{5.28a}$$

From (5.28), using program 2 in Appendix B, we can verify that $s = 5.178$ satisfies (5.28a). Hence, the optimum integer number of high-usage trunks is $s = 5$.

The assumption of ATC = 0.83 erl corresponds to that for 24 trunks used at blocking probability $B = 0.01$ with random traffic 15.29 erl. This assumption has been widely used since this size of trunk group is common in practice. However, non-randomness of overflow traffic should be accounted for providing a given GOS (grade of service). This will be described next.

5.3.2 Given Blocking Probability

We shall outline the optimum design to provide a given blocking probability B by applying the ERT [32]. Using the same notation as in Section 5.2, the ERT is formulated as

$$b^* = b(s,a) + a_1 = b(s^*,a^*), \quad v^* = v(s,a) + a_1 = v(s^*,a^*) \tag{5.29}$$

$$B = \frac{b(s^* + s_1, a^*)}{b(s^*, a^*)} \tag{5.30}$$

where $b(\cdot,\cdot)$ and $v(\cdot,\cdot)$ are defined in (5.10).

The relative system cost measured in c_0, is given by

$$f = s + ks_1. \tag{5.31}$$

Thus, the optimum design for minimizing the system cost with a given blocking probability B in the alternative route, reduces to a *non-linear programing problem* to minimize the *objective function* f in (5.31), subject to the *constraints*,

$$\begin{aligned} g_1 &= b(s^*,a^*) - b^* = 0 \\ g_2 &= v(s^*,a^*) - v^* = 0 \\ g_3 &= b(s^* + s_1, a^*) - Bb(s^*.a^*) = 0 \end{aligned} \tag{5.32}$$

which are obtained from (5.30).

Introduce the *Lagrangean multipliers* τ_1, τ_2 and τ_3, and set

$$F = f + \tau_1 g_1 + \tau_2 g_2 + \tau_3 g_3. \tag{5.33}$$

Then, from

$$\frac{\partial F}{\partial s} = 0, \quad \frac{\partial F}{\partial s_1} = 0, \quad \frac{\partial F}{\partial s^*} = 0, \quad \frac{\partial F}{\partial a^*} = 0$$

we have a set of equations expressed in the matrix form,

$$\begin{bmatrix} 0 & b_s & v_s & 0 \\ 1 & 0 & 0 & b'_s \\ 0 & b^*_s & v^*_s & b'_s - Bb^*_s \\ 0 & b^*_a & v^*_a & b'_a - Bb^*_a \end{bmatrix} \begin{bmatrix} k \\ \tau_1 \\ \tau_2 \\ \tau_3 \end{bmatrix} = \begin{bmatrix} 1 \\ 0 \\ 0 \\ 0 \end{bmatrix}. \tag{5.34}$$

Solving (5.34) for k, we have the condition for minimizing f under the constraint,

$$k = \left[\frac{b_s(b'_a v^*_s - b'_s v^*_a) - v_s(b'_a b^*_s - b'_s b^*_a)}{b'_s(b^*_a v^*_s - b^*_s v^*_a)} - \frac{b_s}{b'_s} B \right]^{-1} \tag{5.35}$$

where

$$\begin{aligned} b_s &= \partial b(s,a)/\partial s, & v_s &= \partial v(s,a)/\partial s \\ b^*_x &= \partial b(s^*,a^*)/\partial x^*, & v^*_x &= \partial v(s^*,a^*)/\partial x^* \\ b'_x &= \partial b(s_1+s^*,a^*)/\partial x^*, & x &= a, s. \end{aligned}$$

These derivatives are estimated by setting $t = s + 1 - a + b$, from

$$\begin{aligned} \partial b(s,a)/\partial a &= tb/a \\ \partial b(s,a)/\partial s &= -b\Psi_{s+1} \\ \partial v(s,a)/\partial a &= v/t - t(b^2 - v)/a \\ \partial v(s,a)/\partial s &= [b^2(1 + a/t^2) - v]\Psi_{s+1} - ab/t^2 \end{aligned} \tag{5.36}$$

where Ψ_{s+1} is calculated by (B.7) in Appendix B.

[Example 5.5] Figure 5.7 shows a result with the traffic load $a = 5\,\text{erl}$ offered to the high-usage route and the given blocking probability $B = 0.01$.

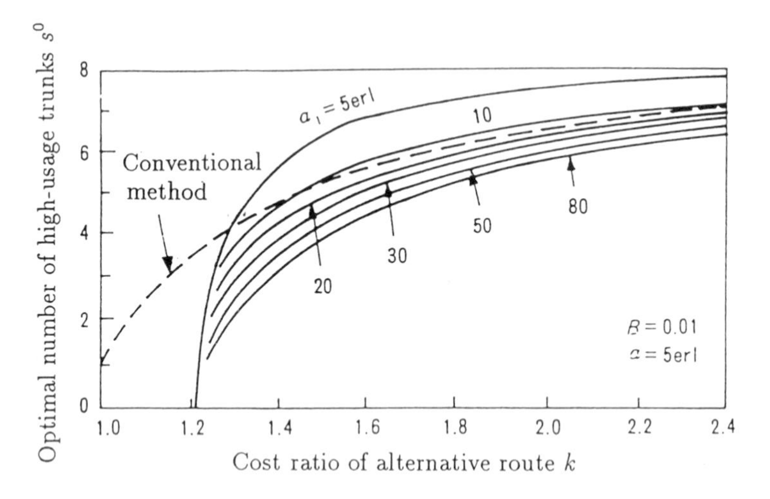

Fig.5.7 Optimum design with given blocking probability

For example, with $k = 1.5$, and $a_1 = 10\,\text{erl}$, we have the optimum number $s^0 = 5$ of high-usage route trunks. This coincides with the result in Example 5.4 by the conventional method, which is also plotted in the figure. It can be seen, however, that the conventional method causes an error for smaller values of k, say $k \le 1.3$, which will be the case for digital transmission and switching.

5.3.3 Trunk Reservation Scheme

As shown in Example 5.3, there is a difference in the individual blocking probabilities B_0 and B_1. To prevent such an unbalance in GOS, the trunk reservation scheme is used. We shall describe the optimum design of the alternative routing system with a trunk reservation scheme.

In the overflow route with a trunk reservation scheme, s_2 out of s_1 trunks are reserved for the background traffic load a_1 for balancing the GOS, as shown in Figure 5.8. Let j and k be the numbers of calls present in the high-usage and overflow routes, respectively, and set $r = s_1 - s_2$. Then, the system is described as follows:

(1) The traffic (a_0) offered to the high-usage route (s_0) is carried by that route if $j < s_0$, or overflows to the overflow route as the *ordinary call* (b) if $j \geq s_0$.

(2) In the overflow route (s_1), the background traffic (a_1) is served as the *priority call* if $k < s_1$, or lost if $k \geq s_1$. The ordinary call (b) is served if $k < r$, or lost if $k \geq r$.

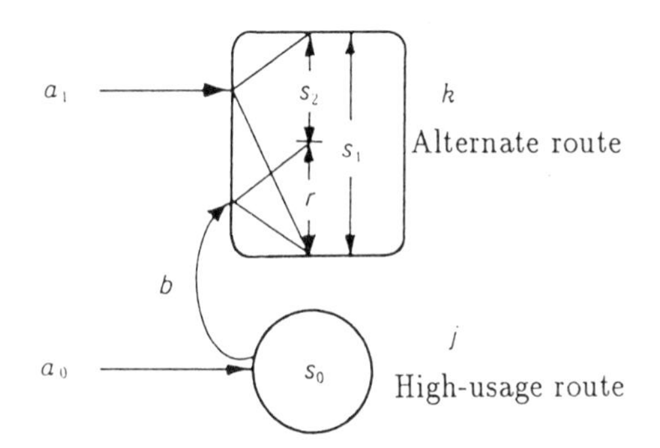

Fig.5.8 Trunk reservation scheme

Denoting the individual blocking probabilities for load a_0 and a_1 by B_0 and B_1, respectively, they are calculated approximately by [34]

$$B_0 = P_0 \frac{s_1!}{a_1^{s_1}} \sum_{j=r}^{s_1} \frac{a_1^j}{j!}, \quad B_1 = P_1 \frac{a_1^{s_1}}{s_1!} \left(\frac{a_2}{a_1}\right)^r \tag{5.37}$$

where

$$\begin{aligned} P_0 &= \left[\frac{a_0 s_1!}{a_1^{s_1}} \sum_{j=r}^{s_1} \frac{a_1^j}{j! b(s_0 + j, a)} + \frac{a_1 s_0!}{a_0^{s_0}} \sum_{j=0}^{s_0} \frac{a_0^j}{j! b(s_1 + j, a)} \right]^{-1} \\ P_1 &= \left[\sum_{j=0}^{r} \frac{a_2^j}{j!} + \left(\frac{a_2}{a_1}\right)^r \sum_{j=r+1}^{s_1} \frac{a_1^j}{j!} \right]^{-1} \end{aligned} \tag{5.38}$$

with $a = a_0 + a_1$, $a_2 = a_1 + b(s_0, a_0)$.

The optimum design for minimizing the system cost with a given individual GOS, is attained by introducing the trunk reservation scheme, in a similar manner as in Subsection 5.3.2 [33].

Given a predetermined blocking probability B, we have the constraint,

$$g_0 = B_0 - B = 0, \quad g_1 = B_1 - B = 0. \tag{5.39}$$

The optimum condition which minimizes the relative system cost,

$$f = s_0 + ks_1 \tag{5.40}$$

subject to the constraint (5.39), is given by

$$k = \frac{B_{11}B_{02} - B_{01}B_{12}}{B_{10}B_{02} - B_{00}B_{12}} \tag{5.41}$$

where k is defined in (5.26), and $B_{ij} \equiv \partial B_i/\partial s_j$, $i, j = 0, 1, 2$, are calculated by using the derivatives of the Erlang B formula [33, p.466]. (See Appendix B.3.)

[Example 5.6] Figure 5.9 shows an example of the optimum design with $a_0 = 5$ erl and $a_1 = 10$ erl, where the predermined GOS is $B = 0.01$.

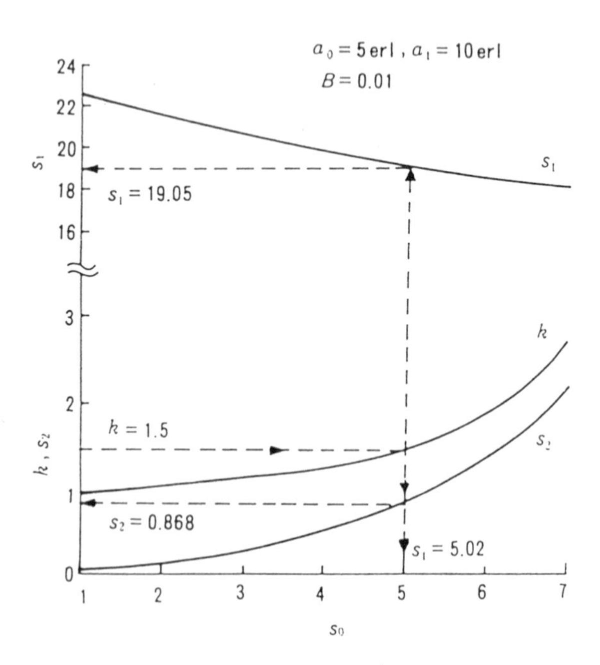

Fig.5.9 Optimum design of trunk reservation scheme

For insttance. given the cost ratio $k = 1.5$, we find the optimum solution $s_0 = 5.02$, $s_1 = 19.05$ and $s_2 = 0.868$ as indicated in the figure. By searching neighbour of the optimum solution, the integer solution satisfying the condition $B_0,\ B_1 \leq 0.01$ is obtained. They are

$$s_0 = 6, \quad s_1 = 19, \quad s_2 = 1.$$

With these values, the blocking probabiliies are calculated as $B_0 = 0.0061$ and $B_1 = 0.0075$ from the approximate formulas (5.37) and (5.38). As will be seen in Table 5.7, the exact values are $B_0 = 0.0076$ and $B_1 = 0.0083$, which satisfy the GOS $B = 0.01$. Thus, it is found that the approximate formulas give fairly good estimate and are useful for practical applications.

5.4 Numerical Analysis of State Equations

5.4.1 State Equation for Alternative Routing

When steady state equations are obtained, but it is difficult to solve them analytically, numerical analysis by iterative calculation is applicable [4, p.158]. As an example, let us show this with the alternative routing system.

In the alternative routing system shown in Figure 5.3, letting P_{jk} be the probability that j calls and k calls exist in the primary and overflow routes, respectively, in the steady state, we have the set of state equations,

$$\begin{aligned}
(a_0 + a_1 + j + k)P_{jk} = &\ (j+1)P_{j+1,k} + (k+1)P_{j,k+1} \\
&+ a_0 P_{j-1,k} + a_1 P_{j,k-1}, \quad 0 \leq j < s_0,\ 0 \leq k < s_1 \\
(a_0 + a_1 + s_0 + k)P_{s_0,k} = &\ (k+1)P_{s_0,k+1} + a_0 P_{s_0-1,k} \\
&+ (a_0 + a_1)P_{s_0,k-1}, \quad j = s_0,\ 0 \leq k < s_1 \\
(a_0 + j + s_1)P_{j,s_1} = &\ (j+1)P_{j+1,s_1} + a_0 P_{j-1,s_1} + a_1 P_{j,s_1-1}, \\
&\quad 0 \leq j < s_0,\ k = s_1 \\
(s_0 + s_1)P_{s_0,s_1} = &\ a_0 P_{s_0-1,s_1} + (a_0 + a_1)P_{s_0,s_1-1}, \quad j = s_0,\ k = s_1
\end{aligned} \tag{5.42}$$

where the offered load and the number of trunks in the high-usage route are re-labeled as a_0 and s_0, respectively. The normalization condition is given by

$$\sum_{j=0}^{s_0} \sum_{k=0}^{s_1} P_{jk} = 1. \tag{5.43}$$

For example, if $s_0 = s_1 = 2$, letting $a = a_0 + a_1$, and re-ordering (5.42), we have the coefficient matrix in Table 5.2. The number of components is $(s_0 + 1) \times$

$(s_1+1) = 9$, all diagonal components are positive, and all off-diagonal components non-positive. This matrix corresponds to a set of 9 equations, but one of them may be derived from others, thus it is not independent. Therefore, from 8 independent equations and the normalization condition, 9 unknown probabilities P_{jk}, $j,k = 0,1,2$, may be obtained. Although one of the equations is thus redundant, by using all of the equations, we can proceed the numerical calculations efficiently. This procedure will be described by using an example.

Table 5.2 Coefficient matrix for alternative routing system

(j,k)	(2,2)	(2,1)	(2,0)	(1,2)	(1,1)	(1,0)	(0,2)	(0,1)	(0,0)
(2,2)	4	$-a$	0	$-a_0$	0	0	0	0	0
(2,1)	-2	$3+a$	$-a$	0	$-a_0$	0	0	0	0
(2,0)	0	-1	$2+a$	$-a$	0	$-a_0$	0	0	0
(1,2)	-2	0	0	$3+a_0$	$-a_1$	0	$-a_0$	0	0
(1,1)	0	-2	0	-2	$2+a$	$-a_1$	0	$-a_0$	0
(1,0)	0	0	-2	0	-1	$1+a$	0	0	$-a_0$
(0,2)	0	0	0	-1	0	0	$2+a_0$	$-a_1$	0
(0,1)	0	0	0	0	-1	0	-2	$1+a$	$-a_1$
(0,0)	0	0	0	0	0	-1	0	-1	a

5.4.2 Gauss-Seidel Iteration

For example, with $a_0 = a_1 = 1\,\text{erl}$ and therefore $a = 2\,\text{erl}$, dividing the raw components by the diagonal ones, from Table 5.2 we obtain Table 5.3.

Table 5.3 Normalized coefficient matrix

(j,k)	(2,2)	(2,1)	(2,0)	(1,2)	(1,1)	(1,0)	(0,2)	(0,1)	(0,0)
(2,2)	1	$-2/4$	0	$-1/4$	0	0	0	0	0
(2,1)	$-2/5$	1	$-2/5$	0	$-1/5$	0	0	0	0
(2,0)	0	$-1/4$	1	$-2/4$	0	$-1/4$	0	0	0
(1,2)	$-2/4$	0	0	1	$-1/4$	0	$-1/4$	0	0
(1,1)	0	$-2/4$	0	$-2/4$	1	$-1/4$	0	$-1/4$	0
(1,0)	0	0	$-2/3$	0	$-1/3$	1	0	0	$-1/3$
(0,2)	0	0	0	$-1/3$	0	0	1	$-1/3$	0
(0,1)	0	0	0	0	$-1/3$	0	$-2/3$	1	$-1/3$
(0,0)	0	0	0	0	0	$-1/2$	0	$-1/2$	1

Suppose a product form solution, and set the initial value,

$$P_{jk}^{(0)} = C\frac{a_0{}^j}{j!}\frac{a_1{}^k}{k!} \tag{5.44}$$

where C is the normalization constant.

The steady state probabilities P_{jk}, $j, k = 0, 1, 2$, are calculated by the following algorithm:

[Algorithm 5.1]

(1) Introduce the weighting factor w, and calculate

$$P_{2,2}^{(1)} = w[(2/4)P_{2,1}^{(0)} + (1/4)P_{1,2}^{(0)}] + (1-w)P_{2,2}^{(0)}. \tag{5.45}$$

(2) Successively calculate

$$\begin{aligned} P_{2,1}^{(1)} &= w[(2/5)P_{2,2}^{(1)} + (2/5)P_{2,0}^{(0)} + (1/5)P_{1,1}^{(0)}] + (1-w)P_{2,1}^{(0)} \\ P_{2,0}^{(1)} &\doteq w[(1/4)P_{2,1}^{(1)} + (2/4)P_{1,2}^{(0)} + (1/4)P_{1,0}^{(0)}] + (1-w)P_{2,0}^{(0)} \\ &\vdots \\ P_{0,0}^{(1)} &= w\left[(1/2)P_{1,0}^{(1)} + (1/2)P_{0,1}^{(1)}\right] - (1-w)P_{0,0}^{(0)}. \end{aligned} \tag{5.46}$$

(3) Normalize as

$$C = \left[\sum_{j=0}^{s_0}\sum_{k=0}^{s_1} P_{jk}^{(1)}\right]^{-1}, \quad CP_{jk}^{(1)} \to P_{jk}^{(1)}. \tag{5.47}$$

This may be done after Step (5).

(4) Repeat (1) to (3) to calculate $P_{jk}^{(2)}$, $P_{jk}^{(3)}, \cdots$.

(5) Stop the iteration if

$$\sum_{j=0}^{s_0}\sum_{k=0}^{s_1}\left|P_{jk}^{(n)} - P_{jk}^{(n-1)}\right| < \epsilon \tag{5.48}$$

where ϵ is the predetermined error.

This method is called *over-relaxation* if $w > 1$, and in particular *Gauss-Seidel iteration* if $w = 1$. From practical experiences, it is known that $w \doteq 1.3$ speeds up convergence.

Table 5.4 shows a result of the iteration which is completed at $n = 6$.

Table 5.4 Example of iteration, $w = 1.3$, $\epsilon = 0.001$

$P_{jk}^{(n)}$	n=0	1	2	3	4	5	6
(2,2)	0.0500	0.0617	0.0664	0.0637	0.0634	0.0638	0.0638
(2,1)	0.0800	0.0874	0.0795	0.0800	0.0805	0.0806	0.0805
(2,0)	0.0800	0.0546	0.0544	0.0557	0.0560	0.0557	0.0557
(1,2)	0.0800	0.0906	0.0961	0.0939	0.0940	0.0941	0.0941
(1,1)	0.1600	0.1681	0.1615	0.1636	0.1636	0.1635	0.1634
(1,0)	0.1600	0.1401	0.1407	0.1431	0.1425	0.1424	0.1424
(0,2)	0.0800	0.0816	0.0871	0.0853	0.0854	0.0854	0.0855
(0,1)	0.1600	0.1635	0.1622	0.1622	0.1623	0.1622	0.1622
(0,0)	0.1600	0.1525	0.1520	0.1524	0.1524	0.1523	0.1523

Using P_{jk} thus obtained, the individual blocking probabilities B_0' and B_1 for the overflow traffic (b) and background traffic (a_1), respectively, are calculated by

$$B_0' = \frac{P_{s_0,s_1}}{\sum_{k=0}^{s_1} P_{s_0,k}}, \quad B_1 = \sum_{j=0}^{s_0} P_{j,s_1}. \tag{5.49}$$

We have the average blocking probability B of the alternative route, and overall blocking probability B_0 for the offered traffic load a_0 to the high-usage route as

$$B = \frac{b_0 B_0' + a_1 B_1}{b_0 + a_1}, \quad B_0 = \frac{b_0 B_0'}{a_0} \tag{5.50}$$

where $b_0 = b(s_0, a_0)$.

[Example 5.7] Table 5.5 shows results for various traffic conditions, which provide the exact values referred in the previous examples.

Table 5.5 Examples of numerical analysis (exact solution)

a_0	s_0	a_1	s_1	B_0	B_1	B
5	5	10	20	0.0132	0.0079	0.0085
15	10	15	30	0.0294	0.0193	0.0222
20	20	20	30	0.0752	0.0358	0.0412

5.4.3 Solution for Trunk Reservation Scheme

In the system with a trunk reservation scheme in Figure 5.8, using similar notation to (5.42), we have the system of steady state equations,

$$\begin{aligned}
(j+k+a)P_{jk} &= a_0 P_{j-1,k} + a_1 P_{j,k-1} + (j+1)P_{j+1,k} + (k+1)P_{j,k+1}, \\
&\qquad 0 \le j < s_0,\ 0 \le k < s_1 \\
(s_0+k+a)P_{s_0,k} &= a_0 P_{s_0-1,k} + a P_{s_0,k-1} + (k+1)P_{s_0,k+1}, \\
&\qquad j = s_0,\ 0 \le k < r \\
(s_0+k+a_1)P_{s_0,k} &= a_0 P_{s_0-1,k} + [a_0\sigma_{kr} + a_1]P_{s_0,k-1} + (k+1)P_{s_0,k+1}, \\
&\qquad j = s_0,\ r \le k < s_1 \\
(j+s_1+a_0)P_{j,s_1} &= a_0 P_{j-1,s_1} + a_1 P_{j,s_1-1} + (j+1)P_{j+1,s_1}, \\
&\qquad 0 \le j < s_0,\ k = s_1 \\
(s_0+s_1)P_{s_0,s_1} &= a_0 P_{s_0-1,s_1} + [a_0\sigma_{r,s_1} + a_1]P_{s_0,s_1-1}, \\
&\qquad j = s_0,\ k = s_1
\end{aligned} \tag{5.51}$$

where $a = a_0 + a_1$, $\sigma_{xy} = 1$ if $x = y$, and $\sigma_{xy} = 0$ if $x \neq y$.

In a similar manner as in Subsection 5.4.2, we can calculate P_{jk}, $j = 0, 1, \cdots, s_0$, $k = 0, 1, \cdots, s_1$. The individual blocking probabilities B_0 and B_1 for the high-usage route traffic load a_0 and background traffic load a_1 in the overflow route, respectively, are calculated by

$$B_0 = \sum_{k=r}^{s_1} P_{s_0,k}, \quad B_1 = \sum_{j=0}^{s_0} P_{j,s_1}. \tag{5.52}$$

[Example 5.8] With $s_0 = s_2 = r = 1$ and $s_1 = 2$, the coefficient matrix is given in Table 5.6. This will be used in Exercise [4] in this chapter.

Table 5.6 Coefficient matrix for trunk reservation scheme

(j,k)	(1,2)	(1,1)	(1,0)	(0,2)	(0,1)	(0,0)
(1,2)	3	$-a_1$	0	$-a_0$	0	0
(1,1)	-2	$2+a_0$	$-a$	0	$-a_0$	0
(1,0)	0	-1	$1+a$	0	0	$-a_0$
(0,2)	-1	0	0	$2+a_0$	$-a_1$	0
(0,1)	0	-1	0	-2	$1+a$	$-a_1$
(0,0)	0	0	-1	0	-1	a

Table 5.7 shows calculated results for various traffic conditions.

Table 5.7 Numerical solution for trunk reservation (Exact)

a_0	s_0	a_1	s_1	s_2	B_0	B_1	B
5	6	10	19	1	0.0076	0.0083	0.0110
10	10	10	20	2	0.0202	0.0084	0.0236
15	10	15	30	5	0.0746	0.0029	0.0550

Exercises

[1] In the overflow system from M/M/1(0) with offered load a and mean service time h, verify

(1) The mean b, and index of variance z of the overflow calls, are given by

$$b = \frac{a^2}{1+a}, \quad z = \frac{2+4a+a^2}{(1+a)(2+a)}. \tag{5.53}$$

(2) The mean m and SCV $C_a{}^2$ of the inter-overflow time, are given by

$$m = \frac{h}{b}, \quad C_a{}^2 = \frac{a}{(1+a)^2}. \tag{5.54}$$

[2] Consider an alternative routing system with random traffic load $a_0 = 10$ erl offered to $s_0 = 10$ trunks in high-usage route, and background random traffic $a_1 = 20$ erl in the overflow route.

(1) Design s_1 for providing the average blocking probability $B \leq 0.01$ in the overflow route, by using the ERT.

(2) Calculate the individual blocking probabilities B_0 and B_1 for a_0 and a_1, by using the GI approximation, and B obtained in (1).

[3] In the alternative routing system in Example 5.4, calculate s_1 (real) for various s_0 (integer), given $B = 0.01$. Then,

(1) Find s_0 to minimize the cost $f = s + ks_1$, with $k = 1.5$.

(2) Examine the error of the conventional method, with $k = 1.1$.

[4] In the alternative routing system with trunk reservation scheme, let $a_0 = a_1 = 1$ erl, $s_0 = s_2 = 1$ and $s_1 = 2$.

(1) Calculate the individual blocking probabilities B_0 and B_1 for a_0 and a_1, using (5.37).

(2) Obtain the exact values by numerical analysis using Table 5.6, and examine the accuracy of the approximation in (1).

Chapter 6

ADVANCED TELETRAFFIC MODELS

As tools for analyzing advanced teletraffic systems, this chapter introduces the main results of the modified diffusion approximation, and matrix analytic method. As examples for latest teletraffic systems, performance analyses of the packet multiplexer system in the ISDN and ATM, as well as token ring and CSMA/CD LANs are also described.

6.1 Renewal Input Multi-Server Model

6.1.1 Diffusion Model

The exact analytical solution for a GI/G/s, renewal input multi-server delay system with general service time, has not yet been obtained. So far, various approaches have been made on this topic [8], and, as an example, this section describes the *modified diffusion approximation* (MDA), which achieves fairly good accuracy in a relatively simple calculation.

In a GI/G/s, if the number (state) of calls existing in the system, which is actually a non-negative integer, is regarded as a real number, it constitutes a continuous-time continuous-state Markov process known as the *diffusion process.* This is called *diffusion approximation.*

Letting $X(t)$ be a real number approximating the number of calls present in the system at time t in the steady state, and $f(x,t)$ its density function, where

$$f(x,t)dx = P\{x < X(t) \leq x + dx\}.$$

Then, we have the diffusion equation [9],

$$\frac{\alpha}{2}\frac{\partial^2}{\partial x^2}f(x,t) - \beta\frac{\partial}{\partial x}f(x,t) - \frac{\partial}{\partial t}f(x,t) = 0. \tag{6.1}$$

As shown in Figure 6.1, setting an appropriate boundary condition (known as the Markov reflecting boundary) at $x = 0$ to keep $X(t) \geq 0$, and solving (6.1), we have

$$\begin{aligned}
f_0(x) &= \frac{2\lambda}{\gamma_1\lambda_1}P_0[\exp(\gamma_1 x) - 1], && 0 < x \leq 1 \\
f_1(x) &= f_0(1)\left(\frac{\alpha_x}{\alpha_1}\right)^{\eta}\exp\left[-\frac{2(x-1)}{K_s{}^2}\right], && 1 < x \leq s \\
f_s(x) &= f_1(s)\exp[\gamma_s(x-s)], && s < x \leq \infty
\end{aligned} \tag{6.2}$$

where λ is the arrival rate, P_0 the probability that the system is empty at an arbitrary time in the steady state, and

$$\alpha_x = \lambda K_a{}^2 + x\mu K_s{}^2, \quad \beta_x = \lambda - x\mu$$
$$\gamma_x = 2\beta_x/\alpha_x, \quad \eta = 2a(K_a{}^2 + K_s{}^2)/K_s{}^4 - 1$$

with $\mu^{-1} = h$ being the mean service time, and $a = \lambda h$ the offered load.

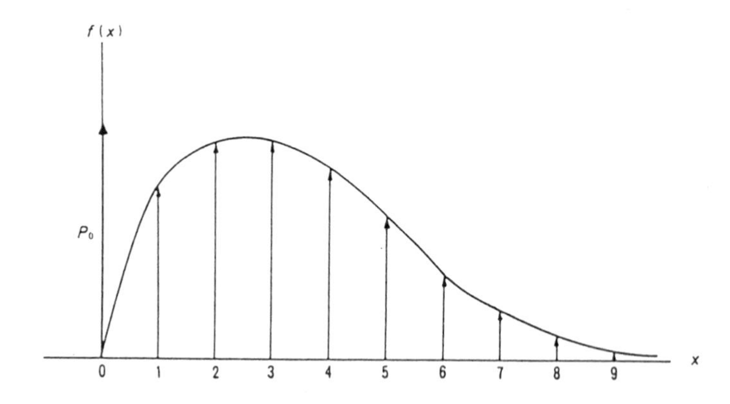

Fig.6.1 Diffusion model

In conventional diffusion approximation, the parameters $K_a{}^2$ and $K_s{}^2$ are set equal to the squared coefficients of variation (SCVs) of interarrival time and service time, respectively, where the *central limit theorem* (See Appendix C.4.3.) is applied assuming a heavy load condition. Consequently, this results in a poor approximation under light load conditions. Therefore, in the *modified diffusion approximation*, $K_a{}^2$ and $K_s{}^2$ are determined as follows to improve the accuracy [35].

6.1.2 Modified Diffusion Approximation

The parameters ${K_a}^2$ and ${K_s}^2$ are determined by comparing the diffusion approximation with the exact solutions for M/G/1 and GI/M/1. With $s = 1$, since the probability that the server is busy, is equal to the carried load a (from Property (3) in Subsection 1.2.1), we have $P_0 = 1 - a$, and from (6.2) the mean number of calls in the system,

$$\begin{aligned} N &= \int_{0+}^{1} x f_0(x) dx + \int_{1+}^{\infty} x f_s(x) dx \\ &= a \left[\frac{1}{2} + \frac{a{K_a}^2 + {K_s}^2}{2(1-a)} \right]. \end{aligned} \tag{6.3}$$

Hence, from the Little formula, the mean waiting time is given by

$$W = \frac{N}{\lambda} - h = \left[\frac{a{K_a}^2 + {K_s}^2}{2(1-a)} - 1 \right] \frac{h}{2}. \tag{6.4}$$

For M/G/1, from (3.20) we have the exact solution,

$$W = \frac{\rho}{1-\rho} \frac{1 + {C_s}^2}{2} h \tag{6.4a}$$

where ${C_s}^2$ is the SCV of service time, and $\rho = a/s$ the utilization factor. By setting ${K_a}^2 = 1$ for M/G/1, and comparing (6.4) with (6.4a), the parameter ${K_s}^2$ is determined by

$${K_s}^2 = 1 + \rho({C_s}^2 - 1). \tag{6.5}$$

Next, for GI/M/1, from (3.84) we have the exact solution,

$$W = \frac{\omega}{1-\omega} h \tag{6.6}$$

where ω is the generalized occupancy. From (3.80) ω is given by the root $(0 < \omega < 1)$ of the equation,

$$\omega = \alpha([1-\omega]s\mu) \tag{6.7}$$

where $\alpha(\theta)$ is the LST of the interarrival time distribution. Noting that ${K_s}^2 = 1$ in (6.5) with ${C_s}^2 = 1$ for the exponential service time, and comparing (6.4) and (6.6), the parameter ${K_a}^2$ is determined by

$${K_a}^2 = \left[(1-\rho) \frac{1+\omega}{1-\omega} - 1 \right] \frac{1}{\rho}. \tag{6.8}$$

Although the *modified parameters* ${K_a}^2$ and ${K_s}^2$ have been derived for $s = 1$, they do not include s explicitly, and hence we shall extend the approximation to the multi-server system.

6.1.3 GI/G/s

For a GI/G/s, let P_j, $j = 1, 2, \cdots$, be the probability that j (integer) calls exist in the system at an arbitrary time in the steady state, and set

$$\begin{aligned} P_j &= f_1(j), \quad j = 1, 2, \cdots, s \\ P_j &= f_s(j), \quad j = s+1, s+2, \cdots \end{aligned} \tag{6.9}$$

where $f_i(\cdot)$ and $f_s(\cdot)$ are given in (6.2). Then, using (6.2) in (6.9), we have

$$P_j = \begin{cases} \sigma_j P_0, & j = 1, 2, \cdots, s \\ P_s \omega_s^{j-s}, & j = s+1, s+2, \cdots \end{cases} \tag{6.10}$$

$$P_0 = \begin{cases} 1 - a, & s = 1 \\ \left(1 + \sum_{j=1}^{s-1} \sigma_j + \dfrac{\sigma_s}{1 - \omega_s}\right)^{-1}, & s = 2, 3, \cdots \end{cases} \tag{6.11}$$

where

$$\begin{aligned} \sigma_j &= \frac{\lambda(\omega_1 - 1)}{\beta_1} \left(\frac{\alpha_j}{\alpha_1}\right)^{\eta} \exp\left[-\frac{2(j-1)}{K_s^{\,2}}\right], \quad j = 1, 2, \cdots, s \\ \omega_j &= \exp(-\gamma_j), \qquad j = 1, s. \end{aligned} \tag{6.12}$$

Denoting the probability that all s servers are busy at an arbitrary instant by $\hat{M}(0)$, and the mean number of waiting calls by L, we have

$$\begin{aligned} \hat{M}(0) &= \sum_{j=s}^{\infty} P_j = P_s \frac{1}{1 - \omega_s} \\ L &= \sum_{j=s}^{\infty} (j - s) P_j = \hat{M}(0) \frac{\omega_s}{1 - \omega_s}. \end{aligned} \tag{6.13}$$

From the Little formula, the mean waiting time W is given by

$$W = \hat{M}(0) \frac{\omega_s}{1 - \omega_s} \frac{h}{a}. \tag{6.14}$$

[Example 6.1] Table 6.1 shows numerical examples of the modified diffusion approximation (MDA) compared with the exact solution or simulation result. It can be seen that the MDA provides fairly good accuracy with a relatively simple calculation, which needs less than 1/100 computer time compared with the simulation.

Table 6.1 Normalized Mean Waiting Time, W/h

Model	a[erl]	s	$C_a{}^2$	$C_s{}^2$	MDA	Exact	Formula
$E_2/H_2/1$	0.5	1	0.5	5	2.6447	2.5909	(3.113)
$H_2/D/1$	0.6	1	10	0	5.8622	5.0262	(3.108)
$M/M/s$	2	3	1	1	0.4458	0.4444	(2.38)
$E_2/M/s$	2	4	0.5	1	0.0375	0.0413	(3.84)
$M/D/s$	3	5	1	0	0.0471	0.0661	(3.58)
$H_2/H_2/s$	4	5	5	5	3.5264	3.10±0.38	Simulation
$H_2/E_2/s$	5	10	10	0.5	0.1248	0.15±0.08	Simulation

± indicates 95% confidence interval.

6.2 PH-MRP Input Models

6.2.1 Phase-Type Markov Renewal Process

In order to analyse sophisticated systems with non-renewal input, we shall introduce the *phase-type Markov renewal process* (PH-MRP) [38][39][40]. To do this, first we need to explain the *phase-type* (PH) *distribution*.

In a continuous-time Markov chain with r transient states and a single $(r+1)$st absorbing state, (See Appendix C.5.) suppose that upon entering the absorbing state, the process instantaneously jumps to transient state j, $j = 1, 2, \cdots, r$, with probability α_j. The *PH distribution* is defined as the inter-visit time distribution to the absorbing state, and characterized by $(\boldsymbol{\alpha}, T)$, where T is the *transition rate matrix* among the transient states, which is an irreducible $r \times r$ matrix. The row vector $\boldsymbol{\alpha}$ with component α_j is called the *initial probability vector*. The column vector $\boldsymbol{T}^\circ$ defined by

$$\boldsymbol{T}^\circ = -T\boldsymbol{e} \tag{6.15}$$

represents the transition rate from transient states to the absorbing state, where $\boldsymbol{e}$ is the *unit column vector* with all components equal to 1. The PH distribution is said to be in *phase* j if the underlying Markov process is in state j. The PH distribution includes hyper-exponential (H_n), Erlangian (E_k), exponential (M) distributions, etc. as the special cases. A renewal process with an interarrival time of PH distribution is called a *phase-type renewal process* (PH-RP).

Next, modify the Markov chain above to have n absorbing states, with jumping probability α_{ij} from absorbing state i, $i = r+1, r+2, \cdots, r+n$, to transient state $j, j = 1, 2, \cdots, r$. Then, the successive visits to the absorbing states constitute the PH-MRP, in which inter-visit times follow PH distributions not identical in general and correlated each other. The PH-MRP is said to be in phase j if the underlying Markov chain is so. The PH-MRP is characterized by *representation* (α, T, T°). The

$n \times r$ matrix α with components α_{ij}, and the $r \times n$ matrix T° are the extensions of the corresponding vectors above, and we have the relation,

$$T^\circ e = -Te. \tag{6.16}$$

The PH-MRP includes the PH-RP as a special case, and can be used for representing versatile renewal and non-renewal processes appearing in modern teletraffic systems, such as the ATM for BISDN. The arrival rate (averaging each PH distribution) is given by

$$\lambda = \pi T^\circ e \tag{6.17}$$

where π is the *stationary probability vector* of $T + T^\circ\alpha$ satisfying

$$\pi(T + T^\circ\alpha) = \mathbf{0}, \quad \pi e = 1. \tag{6.18}$$

6.2.2 PH-MRP/M/$s(m)$

As an example, let us analyse PH-MRP/M/$s(m)$, in which calls arrive in PH-MRP with representation (α, T, T°), and require exponential service time with mean μ^{-1}. If a call finds all the servers busy upon arrival, it waits in a buffer of capacity m for the FIFO service, but if the buffer is full, it is lost.

Denoting the number of calls existing in the system at an arbitrary instant in the steady state by i, and the phase of the PH-MRP at that instant by l, the state space of the model is given by

$$E = \{(i, l);\ 0 \le i \le s + m, \quad 1 \le l \le r\}. \tag{6.19}$$

Furthermore, the state space is divided into a set of levels defined by

$$\text{Level} : \boldsymbol{i} = \{(i, 1),\ (i, 2),\ \cdots,\ (i, r)\}, \quad 0 \le i \le s + m.$$

Then, we can characterize the model as a *quasi birth-death* (QBD) *process* [36], with the infinitesimal generator,

$$Q = \begin{bmatrix} A_0 & B_0 & & & & \mathbf{0} \\ D_1 & A_1 & B_1 & & & \\ & D_2 & A_2 & B_2 & & \\ & & \ddots & \ddots & \ddots & \\ & & & D_{s+m-1} & A_{s+m-1} & B_{s+m-1} \\ \mathbf{0} & & & & D_{s+m} & A_{s+m} \end{bmatrix}. \tag{6.20}$$

The QBD is an extension of the B-D process described in Subsection 2.3.1 (See Appendix C.5.4.), for which components are extended to matrix form, which are given as follows:

The matrix B_i represents the birth-rate in level $\boldsymbol{i}$. As mentioned, since a visit to an absorbing state and jumping to a transient state means a call arrival, we have

$$B_i = T^\circ \alpha, \quad 0 \leq i \leq s+m-1. \tag{6.21}$$

The matrix D_i represents the death rate in level i. Since the service time is exponentially distributed independently of the phase, we have

$$D_i = \begin{cases} i\mu I_r, & 1 \leq i \leq s-1 \\ s\mu I_r, & s \leq i \leq s+m \end{cases} \tag{6.22}$$

where I_r is the *identity matrix* of size $r \times r$ with diagonal components equal to 1.

The matrix A_i represents the rate staying in level i without arrival or termination. The matrix T represents the rate of the process moving among the transient states, but not visiting the absorbing states, that is of no arrival occurs. Since the termination rate is given in (6.22), we have

$$A_i = \begin{cases} T - i\mu I_r, & 0 \leq i \leq s-1 \\ T - s\mu I_r, & s \leq i \leq s+m-1 \\ T - s\mu I_r + T^\circ \alpha, & i = s+m. \end{cases} \tag{6.23}$$

Note that the last equation includes the rate $T^\circ \alpha$ of arriving calls when $i = s+m$, which are lost and do not cause the state change.

Let $\boldsymbol{p}$ be the steady state probability vector, and partition it into blocks such that

$$\boldsymbol{p} = (\boldsymbol{p}_0, \boldsymbol{p}_1, \cdots, \boldsymbol{p}_s, \boldsymbol{p}_{s+1}, \cdots, \boldsymbol{p}_{s+m}) \tag{6.24}$$

where

$$\boldsymbol{p}_i = (p_i(1), p_i(2), \cdots, p_i(r)) \tag{6.25}$$

with $p_i(l)$ being the probability that i calls exist when the system is in phase l. Then, the steady state equation $\boldsymbol{p}Q = \boldsymbol{0}$ yields

$$\begin{aligned} &\boldsymbol{p}_0 T + \mu \boldsymbol{p}_1 = \boldsymbol{0} \\ &\boldsymbol{p}_{i-1} T^\circ \alpha + \boldsymbol{p}_i (T - i\mu I_r) + (i+1)\mu \boldsymbol{p}_{i+1} = \boldsymbol{0}, \quad 1 \leq i \leq s-1 \\ &\boldsymbol{p}_{i-1} T^\circ \alpha + \boldsymbol{p}_i (T - s\mu I_r) + s\mu \boldsymbol{p}_{i+1} = \boldsymbol{0}, \quad s \leq i \leq s+m-1 \\ &\boldsymbol{p}_{s+m-1} T^\circ \alpha + \boldsymbol{p}_{s+m}(T + T^\circ \alpha - s\mu I_r) = \boldsymbol{0}. \end{aligned} \tag{6.26}$$

From the normalization condition $\boldsymbol{pe} = 1$, we have

$$\sum_{i=0}^{s+m} \boldsymbol{p}_i \boldsymbol{e} = 1. \tag{6.27}$$

Solving (6.26), we have the recurrence formula,

$$\begin{aligned} p_i &= (i+1)\mu p_{i+1} C_i^{-1}, && 0 \le i \le s-1 \\ p_i &= s\mu p_{i+1} C_i^{-1}, && s \le i \le s+m-1;\ m \ge 1 \\ p_{s+m} &= p_{s+m} T^\circ \alpha C_{s+m}^{-1}, && m \ge 0 \end{aligned} \tag{6.28}$$

where

$$C_i = \begin{cases} i\mu I_r - T - i\mu C_{i-1}^{-1} T^\circ \alpha, & 0 \le i \le s-1 \\ s\mu I_r - T - s\mu C_{i-1}^{-1} T^\circ \alpha, & s \le i \le s+m. \end{cases} \tag{6.29}$$

Since calls arriving when $i = s+m$ are lost, and their arrival rate is $p_{s+m} T^\circ \alpha e$, the blocking probability is given by

$$B = \lambda^{-1} p_{s+m} T^\circ \alpha e \tag{6.30}$$

where λ is given in (6.17). Using the Little formula, the mean waiting time is given by

$$W = \lambda^{-1} \sum_{i=s+1}^{s+m} (i-s) p_i e. \tag{6.31}$$

[Example 6.2] PH-MRP/M/s(0)

Setting $m = 0$, we have the loss system with PH-MRP input. For example, if we are given

$$\alpha = \begin{bmatrix} 1 & 0 \\ 0 & 1 \end{bmatrix}, \quad T = \begin{bmatrix} -2 & 1 \\ 1/2 & -1 \end{bmatrix}, \quad T^\circ = \begin{bmatrix} 1 & 0 \\ 0 & 1/2 \end{bmatrix}$$

$s = 2$ and $h = \mu^{-1} = 1\,\text{sec}$, from (6.29)

$$C_0^{-1} = \begin{bmatrix} 2/3 & 2/3 \\ 1/3 & 4/3 \end{bmatrix}, \quad C_1^{-1} = \begin{bmatrix} 2/3 & 2/3 \\ 5/12 & 7/6 \end{bmatrix}, \quad C_2^{-1} = \begin{bmatrix} 11/16 & 5/8 \\ 1/2 & 1 \end{bmatrix}.$$

From the last equation in (6.28), we have

$$p_2 = p_2 \begin{bmatrix} 11/16 & 5/8 \\ 1/4 & 1/2 \end{bmatrix}.$$

For example, setting $p_i' = p_i / p_2(2)$ and recursively solving (6.28), we get

$$p_2' = (4/5, 1), \quad p_1' = (19/10, 17/5), \quad p_0' = (12/5, 29/5).$$

Normalizing by (6.27), we get $p_2 = (8/153, 10/153)$. From (6.17) and (6.18), we have $\pi = (1/3, 2/3)$ and $\lambda = (2/3)/\text{sec}$. Hence, from (6.30) we have

$$B = \lambda^{-1} p_2 T^\circ \alpha e = 13/102 = 0.1275.$$

Since (6.20) is given in a unified form, it is applicable to a general PH-MRP input model if the A_i, B_i and D_i are defined appropriately. In fact, it has been successfully applied for analysing sophisticated teletraffic models. For details, refer to [39][40]. An example is shown below.

6.2.3 $\mathbf{M \overleftarrow{+} PH\text{-}MRP/M_1,M_2/}s(\infty,0)$ $\mathbf{PPP}(n)$

Consider the multi-server mixed delay and non-delay system shown in Figure 6.2, in which

(1) Delay input is Poissonian with arrival rate λ_1, and non-delay input is a PH-MRP with representation (α, T, T°) and arrival rate λ_2.

(2) Delay and non-delay calls require exponential service times, with mean μ_1^{-1} and μ_2^{-1}, respectively.

(3) When finding all s servers busy, a delay call waits in an infinite buffer for FIFO service. While a non-delay call preempts a delay call in service, if k $(1 \leq k \leq n)$ delay calls are in service; otherwise it is lost, where n is called the *threshold of preemption.*

This is called *partial preemptive priority* (PPP) scheme with threshold n, and represented by $\mathrm{M} \overleftarrow{+} \mathrm{PH\text{-}MRP/M_1,M_2}/s(\infty,0)$ $\mathrm{PPP}(n)$

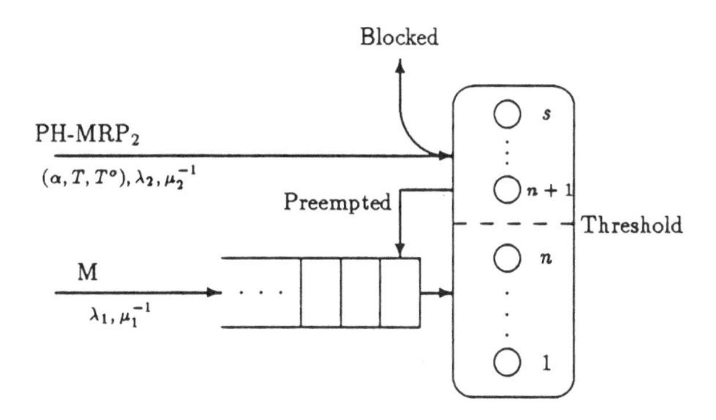

Fig.6.2 $\mathrm{M} \overleftarrow{+} \mathrm{PH\text{-}MRP/M_1,M_2}/s(\infty,0)$ $\mathrm{PPP}(n)$

Define the state space of the model by

$$E = \{(i,j,l)\,;\ i \geq 0,\ 0 \leq j \leq s,\ 1 \leq l \leq r\}$$

where the indices i and j denote the numbers of delay and non-delay calls, respectively, existing in the system at an arbitrary instant in the steady state, and l represents the phase of the PH-MRP at that instant. Partition E into a set of levels and sublevels as

$$\begin{aligned}
\text{Level: } i &= \{(i,0,1),\cdots,(i,0,r),(i,1,1),\cdots,(i,1,r),\cdots,(i,s,1),\cdots,(i,s,r)\} \\
\text{Sublevel: } j &= \{(j,1),\cdots,(j,r)\}.
\end{aligned}$$

Then, we can characterize the model as a QBD with the infinitesimal generator,

$$Q = \begin{bmatrix} A_0 & B_0 & & & & & & & \\ D_1 & A_1 & B_1 & & & & & 0 & \\ & D_2 & A_2 & B_2 & & & & & \\ & & \ddots & \ddots & \ddots & & & & \\ & & & D_{s-1} & A_{s-1} & B_{s-1} & & & \\ & & & & D_s & A_s & B_s & & \\ & 0 & & & & D_s & A_s & B_s & \\ & & & & & & \ddots & \ddots & \ddots \end{bmatrix}. \tag{6.32}$$

The matrix B_i represents the birth rate in level i, and is given by

$$B_i = \begin{array}{c} j \\ 0 \\ \\ \vdots \\ \\ s \end{array} \begin{array}{c} \begin{array}{ccccc} 0 & 1 & \cdots & s\text{-}1 & s \end{array} \\ \begin{bmatrix} \lambda_1 I_r & & & & 0 \\ & \lambda_1 I_r & & & \\ & & \ddots & & \\ & & & \lambda_1 I_r & \\ 0 & & & & \lambda_1 I_r \end{bmatrix} \end{array}. \tag{6.33}$$

The matrix D_i represents the death rate in level i, and is given by

$$D_i = \begin{array}{c} j \\ 0 \\ \vdots \\ s\text{-}i \\ s\text{-}i\text{+}1 \\ \vdots \\ s\text{-}1 \\ s \end{array} \begin{array}{c} \begin{array}{ccccccc} 0 & \cdots & s\text{-}i & s\text{-}i\text{+}1 & \cdots & s\text{-}1 & s \end{array} \\ \begin{bmatrix} i\mu_1 I_r & & & & & & \\ & \ddots & & & & 0 & \\ & & i\mu_1 I_r & & & & \\ & & & (i-1)\mu_1 I_r & & & \\ & & & & \ddots & & \\ & 0 & & & & \mu_1 I_r & \\ & & & & & & 0 \end{bmatrix} \end{array}. \tag{6.34}$$

The matrix A_i represents the rate staying in level i, and is given by

$$A_i =$$

$$\begin{array}{c} j \\ 0 \\ 1 \\ 2 \\ \vdots \\ s\text{-}i_n \\ s\text{-}i_n\text{+}1 \\ \vdots \\ s\text{-}1 \\ s \end{array} \begin{array}{c} \begin{array}{ccccccccc} 0 & 1 & 2 & \cdots & s\text{-}i_n & s\text{-}i_n\text{+}1 & \cdots & s\text{-}1 & s \end{array} \\ \begin{bmatrix} A_i(0) & T^\circ\alpha & & & & & & \\ \mu_2 I_r & A_i(1) & T^\circ\alpha & & & & 0 & \\ & 2\mu_2 I_r & A_i(2) & T^\circ\alpha & & & & \\ & \ddots & \ddots & \ddots & & & & \\ & & (s-i_n-1)\mu_2 I_r & A_i(s-i_n-1) & T^\circ\alpha & & & \\ & & & (s-i_n)\mu_2 I_r & A_i(s-i_n) & 0 & & \\ & & & \ddots & \ddots & \ddots & & \\ & 0 & & & (s-1)\mu_2 I_r & A_i(s-1) & 0 \\ & & & & & s\mu_2 I_r & A_i(s) \end{bmatrix} \end{array}$$

$$(6.35)$$

where $i_n = \min(i, n)$, and

$$A_i(j) = \begin{cases} T - (\lambda_1 + i\mu_1 + j\mu_2)I_r, & 0 \le i \le s;\ 0 \le j < s - i \\ T - [\lambda_1 + (s-j)\mu_1 + j\mu_2]I_r, & n \le i \le s;\ s - i \le j \le s - n \\ T - [\lambda_1 + (s-j)\mu_1 + j\mu_2]I_r + T^\circ \alpha, & 0 \le i \le s;\ s - i_n \le j \le s. \end{cases} \tag{6.36}$$

Partition the stationary probability vector p in levels and sublevels as

$$\begin{aligned} p &= (p_0,\ p_1, \cdots,\ p_s,\ p_{s+1}, \cdots) \\ p_i &= (p_{i0},\ p_{i1},\ \cdots,\ p_{is}) \\ p_{ij} &= (p_{ij}(1),\ p_{ij}(2),\ \cdots,\ p_{ij}(r)) \end{aligned} \tag{6.37}$$

where $p_{ij}(l)$ denotes the stationary probability that i delay calls and j non-delay calls exist in the system at an arbitrary instant, and the PH-MRP is in phase l at that instant. Then, according to the standard matrix-analytic procedure [36], we get the stationary probabilities and the performance measures as follows:

The stationary probability vectors are given by

$$\begin{aligned} p_i &= p_{s-1}R^{i-s+1}, \qquad i \ge s \\ p_i &= p_{s-1}C_i, \qquad 0 \le i \le s-2 \end{aligned} \tag{6.38}$$

where the rate matrix R is the minimal non-negative solution to the *matrix-quadratic equation*,

$$R^2 D_s + RA_s + B_s = 0. \tag{6.39}$$

The matrices C_i, $i = 0, 1, \cdots, s-1$, are given recursively by

$$\begin{aligned} C_{s-1} &= I \\ C_{s-2} &= (-A_{s-1} - RD_s)B_{s-2}{}^{-1} \\ C_{s-i} &= (-C_{s-i+1}A_{s-i+1} - C_{s-i+2}D_{s-i+2})B_{s-i}{}^{-1}, \quad 3 \le i \le s. \end{aligned} \tag{6.40}$$

where $C_1 = R$ if $s = 1$. The boundary probability vector p_{s-1} is the unique solution of the system of equations,

$$\begin{aligned} & p_{s-1}(C_0 A_0 + C_1 D_1) = \mathbf{0} \\ & p_{s-1}\left[\sum_{i=0}^{s-2} C_i + (I - R)^{-1}\right] e = 1. \end{aligned} \tag{6.41}$$

The mean waiting time W for delay traffic, and the blocking probability B for non-delay traffic are given by

$$W = \lambda_1^{-1} \boldsymbol{p}_{s-1} \left[\sum_{i=0}^{s-2} iC_i + (I - R)^{-2} + (s-2)(I - R)^{-1} \right] \boldsymbol{e} - \mu_1^{-1} \tag{6.42}$$

$$B = \lambda_2^{-1} \boldsymbol{p}_{s-1} \left[\sum_{i=0}^{s-1} C_i \Lambda_i + R(I - R)^{-1} \Lambda_s \right] \boldsymbol{e} \tag{6.43}$$

where Λ_i is the *overflow rate matrix* from level $\boldsymbol{i}$, given by

$$\Lambda_i = \begin{array}{c} \\ 0 \\ \vdots \\ s\text{-}i_n\text{-}1 \\ s\text{-}i_n \\ \vdots \\ s \end{array} \begin{array}{c} j \quad 0 \cdots s\text{-}i_n\text{-}1 \quad s\text{-}i_n \quad \cdots \quad s \\ \left[\begin{array}{ccccc} 0 & & & & 0 \\ & \ddots & & & \\ & & 0 & & \\ & & & T^{\circ}\alpha & \\ & & & & \ddots \\ 0 & & & & T^{\circ}\alpha \end{array} \right] \end{array} . \tag{6.44}$$

6.2.4 Special Cases

By specifying the matrices in (6.32), we can deal with various models as special cases. Some examples are shown below.

[Example 6.3] $\mathrm{M}_1 \overleftarrow{+} \mathrm{M}_2/\mathrm{M}_1, \mathrm{M}_2/s(\infty, 0)$

Consider a call acceptance control in the ATM, where the packet switched (delay) data and circuit switched telephone (non-delay) calls are integrated with the PPP scheme and processed by s CLADs (cell assembler-disassembler). Assume that the data packets and telephone calls arrive at Poisson rates λ_1 and λ_2, and hold the CLAD for exponentially distributed time with mean μ_1^{-1} and μ_2^{-1}, respectively. Then, the system is modeled as indicated above. For example, suppose that $\lambda_1 = 60$ packets/sec and $\mu_1^{-1} = 0.125\,\mathrm{sec}$ for data; and $\lambda_2 = 0.06\,\mathrm{calls/sec}$ and $\mu_2^{-1} = 125\,\mathrm{sec}$ for telephone. With $s = 20$ and $n = 6$, from (6.42) and (6.43) we get

Mean waiting time for data packet : $W = 0.2324\,\mathrm{sec}$

Blocking prob. for telephone call : $B = 0.0114$

[Example 6.4] $\mathrm{M} + \mathrm{H}_2/\mathrm{M}/s(\infty, 0)$

The H_2 is the special case of PH-RP characterized by $(\boldsymbol{\alpha}, T)$ with [41]

$$\boldsymbol{\alpha} = (1, 0), \quad T = \begin{bmatrix} \sigma_1 - \xi & -\sigma_1 \\ -\sigma_2 & \sigma_2 \end{bmatrix}, \quad \boldsymbol{T}^{\circ} = \begin{pmatrix} \xi \\ 0 \end{pmatrix} \tag{6.44a}$$

$$\xi = kr_1 + (1-k)r_2, \quad \sigma_1 = k(1-k)(r_1 - r_2)^2/\xi, \quad \sigma_2 = r_1 r_2/\xi. \tag{6.44b}$$

where k, r_1 and r_2 are defined in (4.64a). This PH-RP is also the special case of PH-MRP with representation (α, T, T°), where

$$\alpha = \begin{bmatrix} 1 & 0 \\ 0 & 1 \end{bmatrix}, \quad T = \begin{bmatrix} \sigma_1 - \xi & -\sigma_1 \\ -\sigma_2 & \sigma_2 \end{bmatrix}, \quad T^\circ = \begin{bmatrix} \xi & 0 \\ 0 & 0 \end{bmatrix}. \tag{6.44c}$$

Let us obtain the exact solution for $\mathrm{M}+\mathrm{H}_2/\mathrm{M}/s(\infty, 0)$ in Example 4.8. With the given parameters, $k = 0.9082$, $r_1 = 3.5330$ and $r_2 = 0.3670/\mathrm{sec}$, from (6.44b) we have

$$\xi = 3.2424, \quad \sigma_1 = 0.2577, \quad \sigma_2 = 0.4/\mathrm{sec}.$$

Hence, for $\lambda_1 = \lambda_2 = 2/\mathrm{sec}$, $s = 5$ and $h = \mu^{-1} = 1\,\mathrm{sec}$, from (6.42) and (6.43), we have

$$W = 0.0950\,\mathrm{sec}, \quad B = 0.3810.$$

The corresponding GI approximations in Example 4.8 were $W_2 = 0.0970\,\mathrm{ms}$ and $B_1 = 0.3051$. The exact solutions for various arrival rates have been plotted in Figure 4.12, which demonstrates the accuracy of the GI approximation.

6.3 MMPP Input Model

6.3.1 Markov Modulated Poisson Process

The *Markov modulated Poisson process* (MMPP) is a doubly stochastic Poisson process with arrival rates depending on the phases (states) which constitute a continuous-time Markov chain. (See Appendix C.5.3.) The MMPP is a special case of the PH-MRP, and because of its tractability it is widely used for modeling bursty traffic such as packetized voice in the ATM. The MMPP has been analysed by using the *matrix analytic method* [37], for which the main results useful for applications are summarized here. For details, refer to [37], and [41] where an excellent digest can be found.

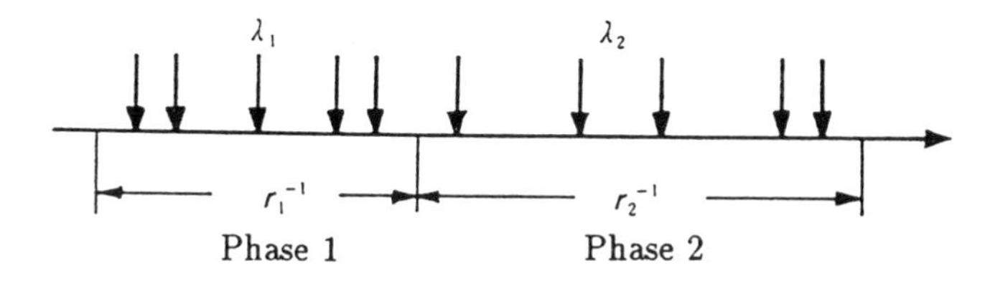

Fig.6.3 Two-phase MMPP

As the simplest case, Figure 6.3 shows the 2-phase MMPP having Poisson arrival rate λ_j in phase j, $j = 1, 2$, which appears alternately with exponentially distributed lifetime with mean r_j^{-1}. This is characterized by (R, Λ) where R is the *infinitesimal generator* of the underlying Markov chain and Λ the *arrival rate matrix*, defined by

$$R = \begin{bmatrix} -r_1 & r_1 \\ r_2 & -r_2 \end{bmatrix}, \quad \Lambda = \begin{bmatrix} \lambda_1 & 0 \\ 0 & \lambda_2 \end{bmatrix}. \tag{6.45}$$

The n-phase MMPP is similarly characterized by (R, Λ) with each matrix of $n \times n$ size. As mentioned, the MMPP is a special case of the PH-MRP, for which parameters are related as

$$\alpha = I, \quad T = R - \Lambda, \quad T^\circ = \Lambda \tag{6.46}$$

where I denotes the *identity matrix* of the appropriate size with diagonal components 1. Although the matrix analysis is applicable for general phase MMPP, the 2-phase MMPP is most tractable and useful for applications, which is often meant without indicating the phase where obvious.

In special cases, the MMPP becomes a renewal process: If $\lambda_1 = \lambda_2 = \lambda$, it reduces to Poisson process with rate λ. If $\lambda_2 = 0$, it becomes one with H_2 (2nd order hyper-exponential distribution) interarrival time, for which parameters indicated with subscript $_H$ are related as follows: (See (3.100).)

MMPP to H_2:

$$\left.\begin{matrix} \lambda_{H1} \\ \lambda_{H2} \end{matrix}\right\} = [\lambda_1 + r_1 + r_2 \pm \sqrt{(\lambda_1 + r_1 + r_2)^2 - 4\lambda r_2}]/2$$
$$k_H = (\lambda_1 - \lambda_{H2})/(\lambda_{H1} - \lambda_{H2}). \tag{6.47}$$

H_2 to MMPP:

$$\lambda_1 = k_H \lambda_{H1} + (1 - k_H)\lambda_{H2}, \quad \lambda_2 = 0$$
$$r_1 = k_H(1 - k_H)(\lambda_{H1} - \lambda_{H2})^2/\lambda_1, \quad r_2 = \lambda_{H1}\lambda_{H2}/\lambda_1. \tag{6.48}$$

[Example 6.5] Consider the superposition of H_2 input and Poisson input. If the H_2 parameters are given by

$$\lambda_{H1} = 0.1, \quad \lambda_{H2} = 0.2/\text{sec}, \quad k = 0.4$$

from (6.48) we have the MMPP parameters,

$$\lambda_1 = 0.16, \quad \lambda_2 = 0, \quad r_1 = 0.015, \quad r_2 = 0.125/\text{sec}.$$

If the Poisson arrival rate is $\lambda_P = 0.2$/sec, since superposition of Poisson is again Poissonian (See Section 1.3.2.), we have Poisson arrival rates λ_1' and λ_2' in respective phases,

$$\lambda_1' = \lambda_1 + \lambda_P = 0.36, \quad \lambda_2' = \lambda_P = 0.2/\text{sec}.$$

Hence, the superposed process, $H_2 + M$, becomes MMPP characterized by (R, Λ) with

$$R = \begin{bmatrix} -0.015 & 0.015 \\ 0.125 & -0.125 \end{bmatrix}, \quad \Lambda = \begin{bmatrix} 0.36 & 0 \\ 0 & 0.2 \end{bmatrix}.$$

6.3.2 MMPP/G/1

Letting $N(t)$ be the number of arrivals in time interval $(0, t]$, we have its probability generating function,

$$g(z,t) = \sum_{i=0}^{\infty} z^j P\{N(t) = j\} = \boldsymbol{\pi} \exp\{[R - (1-z)\Lambda]t\}\boldsymbol{e}. \tag{6.49}$$

where the exponential function for matrix A is defined by

$$\exp(A) \equiv I + A + \frac{A^2}{2!} + \frac{A^3}{3!} + \cdots$$

and $\boldsymbol{\pi}$ is the stationary probability vector satisfying

$$\boldsymbol{\pi} R = 0, \quad \boldsymbol{\pi e} = 1. \tag{6.50}$$

Equation (6.49) is interpreted as an extension, involving R to represent the arrival rate changing, to that of Poisson arrival with rate λ, (See Table C.1 in Appendix C.)

$$g_M(z,t) = \exp[-(1-z)\lambda t]. \tag{6.51}$$

Consider the MMPP/G/1 model with MMPP input characterized by (R, Λ), infinite buffer, FIFO discipline and single server having a general service time distribution.

Define the distribution function $\boldsymbol{W}(t)$ of *virtual waiting time* by a row vector whose components $W_j(t)$ are the waiting time distribution function for the Poisson arrivals in phase j. Then, $\boldsymbol{W}(t)$ satisfies the Volterra integral equation,

$$\boldsymbol{W}(t) = \boldsymbol{P}_0 + \int_0^t \boldsymbol{W}(t-x)[1 - B(x)]dx\,\Lambda - \int_0^t \boldsymbol{W}(t)dx\,R \tag{6.52}$$

where $B(t)$ is the service time distribution function, and $\boldsymbol{P}_0$ is the raw vector whose components are the probabilities that the system is empty in respective phases at an arbitrary time in the steady state. Note that (6.52) is a generalization of the corresponding equation for M/G/1 in (3.29a).

From (6.39) the LST of $\boldsymbol{W}(t)$ is given by

$$w^*(\theta) = \begin{cases} \theta \boldsymbol{P}_0[\theta I + R - (1 - b^*(\theta))\Lambda]^{-1} & \theta > 0 \\ \boldsymbol{\pi} & \theta = 0 \end{cases} \tag{6.53}$$

where $b^*(\theta)$ is the LST of $B(t)$, and $\boldsymbol{\pi}$ is the stationary probability vector satisfying

$$\boldsymbol{\pi} R = \mathbf{0}, \quad \boldsymbol{\pi e} = 1. \tag{6.53a}$$

Equation (6.53) is also a generalization of the LST for M/G/1 in (3.29) which is rewritten as

$$w^*(\theta) = \theta P_0[\theta - (1 - b^*(\theta))\lambda]^{-1} \tag{6.54}$$

where P_0 is the probability of the system being empty.

6.3.3 Mean Waiting Times

The mean virtual waiting time vector is given by

$$\tilde{\boldsymbol{W}} = -\lim_{\theta \to 0} \frac{d\boldsymbol{w}^*(\theta)}{d\theta} = (W_v - 1)\boldsymbol{\pi} + (\boldsymbol{P}_0 + \boldsymbol{\pi}\Lambda h)(\boldsymbol{e\pi} - R)^{-1} \tag{6.55}$$

where W_v is the *mean virtual waiting time* (scalar), and we have the relation,

$$W_v = \tilde{\boldsymbol{W}}\boldsymbol{e} = \frac{2(\boldsymbol{P}_0 + \boldsymbol{\pi}\Lambda)(\boldsymbol{e\pi} - R)^{-1}\tilde{\boldsymbol{\lambda}} h - 2\rho + h^{(2)}\lambda_t}{2(1-\rho)} \tag{6.56}$$

with $\tilde{\boldsymbol{\lambda}} = \Lambda \boldsymbol{e}$, $\lambda_t = \boldsymbol{\pi}\tilde{\boldsymbol{\lambda}}$ being the total arrival rate, h and $h^{(2)}$ the mean and second moment of service time, respectively, and $\rho = \lambda_t h$ the utilization factor. A steady state exists if and only if $\rho < 1$. The *mean actual waiting time* W_a experienced by an arbitrary arrival is given by

$$W_a = \frac{\tilde{\boldsymbol{W}}\tilde{\boldsymbol{\lambda}}}{\lambda_t}. \tag{6.57}$$

Define the *index of waiting time* by

$$u = (W_a - W_v)\lambda_t = (\boldsymbol{P}_0 + \boldsymbol{\pi}\Lambda h)(\boldsymbol{e\pi} - R)^{-1}\tilde{\boldsymbol{\lambda}} - \lambda_t. \tag{6.58}$$

Then, we can express W_v and W_a in terms of u as [43]

$$\begin{aligned} W_v &= W_M + \frac{uh}{1-\rho} \\ W_a &= W_M + \frac{uh}{\rho(1-\rho)} \end{aligned} \tag{6.59}$$

where W_M is the mean waiting time of the M/G/1 with the same arrival rate, and from (3.20) is given by

$$W_M = \frac{\lambda_t h^{(2)}}{2(1-\rho)}. \tag{6.60}$$

The preceding expressions hold for general phase MMPP input. In particular for the 2-phase MMPP input model, we have $\boldsymbol{\pi} = (\pi_1, \pi_2)$,

$$\pi_1 = \frac{r_2}{r_i + r_2}, \quad \pi_2 = \frac{r_1}{r_1 + r_2}, \quad \lambda_t = \frac{\lambda_1 r_2 + \lambda_2 r_1}{r_1 + r_2}. \tag{6.61}$$

$$u = \frac{\lambda_1 - \lambda_2}{(1-\rho)(r_1 + r_2)^2}[r_1 P_{01}(1 - \lambda_2 h) - r_2 P_{02}(1 - \lambda_1 h)] \tag{6.62}$$

where P_{0j}, $j = 1, 2$, are the components of $\boldsymbol{P}_0$, and are calculated as follows:

$$P_{01} = \frac{w(z) - R_1(z) - r_2}{(\lambda_2 - \lambda_1)(1-z)}(1-\rho), \quad P_{02} = 1 - \rho - P_{01} \tag{6.63}$$

with $R_j(z) = \lambda_j(1-z) + r_j$, $j = 1, 2$, $z = b^*(w(z))$, and

$$w(z) = \frac{R_1(z) + R_2(z) + \sqrt{[R_1(z) + R_2(z)]^2 - 4[R_1(z)R_2(z) - r_1 r_2]}}{2} \tag{6.64}$$

which can readily be calculated by iteration.

The mean virtual waiting time in each phase is given by

$$W_j = \pi_j \left(W_v + u \frac{\lambda_j - \lambda_t}{G} \right), \quad j = 1, 2$$
$$G = \sum_{j=1}^{2} \pi_j (\lambda_j - \lambda_t)^2. \tag{6.65}$$

[Example 6.6] Consider an MMPP/D/1 model with the MMPP input in Example 6.5, and fixed service time $h = 2\,\text{sec}$.

From (6.48), we have $\lambda_t = 0.3429/\text{sec}$ and $\rho = 0.6857$. For the fixed service time, we have (See Table C.2 in Appendix C.)

$$b^*(\theta) = \exp(-h\theta).$$

Using this, from (6.62) to (6.64) we have

$$P_{01} = 0.26315, \quad P_{02} = 0.05113, \quad u = 0.01503.$$

The mean waiting times are calculated from (6.60) and (6.59) as

$$W_M = 2.1820\,\text{sec}, \quad W_v = 2.2775\,\text{sec}, \quad W_a = 2.3213\,\text{sec}.$$

6.4 Statistical Packet Multiplexer

6.4.1 Modeling of Packetized Process

In ISDN, multi-media inputs such as voice, data, image, etc., are integrated in the form of packets. In the standard packet switching system (X.25), the packet is in general of variable length. On the other hand, in the ATM for broadband ISDN, it becomes a fixed length cell. In such systems, real-time voice or video is delay-sensitive, whereas data is loss-sensitive, and their performance evaluation is an important problem. Since the packetized process has a bursty nature, the superposed process from multiple sources becomes non-renewal, and analyses using the MMPP have been widely applied.

A model for packetized voice (video) is shown in Figure 6.4, which is often referred to as the *ON-OFF model.* In a single voice source, it is assumed that voice spurt and silence periods are exponentially distributed with mean α^{-1} and β^{-1}, respectively, and packets (cells) are originated in a fixed period T during the voice spurt [42].

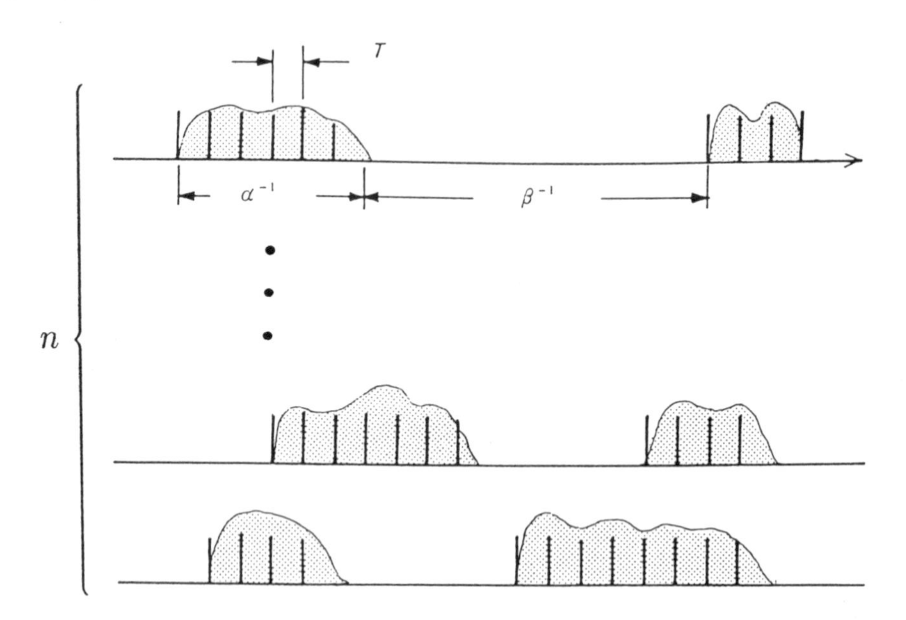

Fig.6.4 ON-OFF model for packetized process

For the arrival process of the single source, the arrival rate λ, SCV (variance/mean^2) $C_a{}^2$, and skewness (third central moment/variance$^{3/2}$) S_k are given by [42]

$$\lambda = \frac{\beta}{T(\alpha+\beta)}$$
$$C_a^2 = \frac{1-(1-\alpha T)^2}{T^2(\alpha+\beta)^2} \tag{6.66}$$
$$S_k = \frac{2\alpha T(\alpha^2T^2 - 3\alpha T + 3)}{[\alpha T(2-\alpha T)]^{3/2}}.$$

Approximating the packet arrival process by H_2 (special case of MMPP) with the first 3 moments matching to (6.66), its LST is given by

$$a^*(\theta) = \frac{k_H\lambda_{H1}}{\theta+\lambda_{H1}} + \frac{(1-k_H)\lambda_{H2}}{\theta+\lambda_{H2}}. \tag{6.67}$$

The parameters are determined from (3.101) in terms of λ, ${C_a}^2$, and S_k as

$$\lambda_{H1} = \lambda(2 - S_kC_a^3 + \sqrt{C})/(3C_a^4 - 2S_kC_a^3 + 1)$$
$$C \equiv (S_k^2 + 18)C_a^6 - 12S_kC_a^5 - 18C_a^4 + 8S_kC_a^3 + 6C_a^2 - 2$$
$$k_H = (C_a^2 - 1)\lambda_{H1}^2/[(C_a^2+1)\lambda_{H1}^2 - 2\lambda(2\lambda_{H1} - \lambda)]$$
$$\lambda_{H2} = \lambda\lambda_{H1}(k_H - 1)/(k_H\lambda - \lambda_{H1}).$$

6.4.2 Indices of Dispersion and Skewness

Letting $N(t)$ be the number of packets arriving in time interval $(0,t]$from the single ON-OFF source, its mean is clearly λt. Denote its variance and third central moment by $V(t)$ and $T(t)$, respectively, and define the *index of variation* $I(t)$, and *index of skewness* $S(t)$ by

$$I(t) \equiv \frac{V(t)}{\lambda t}, \quad S(t) \equiv \frac{T(t)}{\lambda t}. \tag{6.68}$$

Then, we have the relation [43],

$$I(\infty) = {C_a}^2$$
$$I'(0) = \lim_{t\to 0}\frac{d}{dt}I(t) = a(0) - \lambda \tag{6.69}$$
$$S(\infty) = 3{C_a}^4 - S_k{C_a}^3.$$

Letting $a(t)$ be the density function of the packet interarrival time, using the initial value theorem (See Appendix C [T5].) and (6.67), we have

$$\begin{aligned} a(0) &\equiv \lim_{t\to 0} a(t) = \lim_{\theta\to\infty} \theta a^*(\theta) \\ &= k_H\lambda_{H1} + (1-k_H)\lambda_{H2}. \end{aligned} \tag{6.70}$$

Consider the superposition of packet arrival processes from n independent sources, which is not renewal because of the burstiness of packet arrival process. From the additive property of the arrival rate, variance and third central moment, we have the *conservation law* for the indices,

$$I_n(t) = \frac{nV(t)}{n\lambda} = I(t), \quad S_n(t) = \frac{nT(t)}{n\lambda} = S(t). \tag{6.71}$$

That is, the indices for the superposed process are identical to the ones for the single source. On the other hand, the arrival rate becomes $\lambda_n = n\lambda$.

For a general phase MMPP characterized by (Λ, R), from (6.36) we have the Laplace transform (LT) of the generating function $g(t,z)$ of $N(t)$,

$$g^*(\theta, z) = \pi[(1-z)\Lambda - R][I\theta + (1-z)\Lambda - R]^{-1}\mathbf{e}. \tag{6.72}$$

Letting $M_n(t)$ be the nth factorial moment of $N(t)$, its LT is given by

$$M_n^*(\theta) = \lim_{z\to 1} \frac{\partial^n}{\partial z^n} g^*(\theta, z). \tag{6.73}$$

Inverting $M_n^*(\theta)$, and using the relations, $M_1(t) = \lambda_t t$ and

$$\begin{aligned} V(t) &= M_2(t) + \lambda_t t - (\lambda_t t)^2 \\ T(t) &= M_3(t) - V(t)(\lambda_t t - 1) - \lambda_t t(\lambda_t t - 1)[M_2(t) - 2] \end{aligned} \tag{6.74}$$

we get the indices and total arrival rate λ_M for the MMPP [43],

$$\begin{aligned} I_M(\infty) &= 1 + 2\pi\Lambda[\mathbf{e}\pi - (R + \mathbf{e}\pi)^{-1}]\tilde{\lambda}\lambda_t^{-1} \\ I'_M(0) &= \pi\Lambda R^2(R + \mathbf{e}\pi)^{-2}\tilde{\lambda}\lambda_t^{-1} \\ S_M(\infty) &= 1 + 6\pi\Lambda[\mathbf{e}\pi - (R + \mathbf{e}\pi)^{-1}][I + \tilde{\lambda}\pi + (R + \mathbf{e}\pi)^{-1}\tilde{\lambda}\pi \\ &\qquad -\Lambda(R + \mathbf{e}\pi)^{-1}]\tilde{\lambda}\lambda_t^{-1} \\ \lambda_M &= \lambda_t = \pi\tilde{\lambda} \end{aligned} \tag{6.75}$$

where the notation is the same as in Section 6.3.

6.4.3 MMPP Approximation

Let us approximate the superposed packet process by a 2-phase MMPP characterized by (R, Λ) in (6.32). From (6.75), we have the indices and the arrival rate λ_M,

$$\begin{aligned} I_M(\infty) &= 1 + \frac{2r_1r_2(\lambda_1-\lambda_2)^2}{\lambda(r_1+r_2)^3} \\ I'_M(0) &= \frac{r_1r_2(\lambda_1-\lambda_2)^2}{\lambda(r_1+r_2)^2} \\ S_M(\infty) &= 1 + \frac{6r_1r_2(\lambda_1-\lambda_2)^2}{\lambda(r_1+r_2)^3}\left[1+\frac{(r_1-r_2)(\lambda_1-\lambda_2)}{(r_1+r_2)^2}\right] \\ \lambda_M &= \frac{r_1\lambda_2+r_2\lambda_1}{r_1+r_2}. \end{aligned} \tag{6.76}$$

Making the corresponding indices in (6.76) match those in (6.69) and setting $\lambda_M = \lambda_n = n\lambda$, the 2-phase MMPP parameters are explicitly determined by

$$\left.\begin{matrix} r_1 \\ r_2 \end{matrix}\right\} = D\left(1 \pm \frac{1}{\sqrt{1+\lambda_n E}}\right) \qquad \left.\begin{matrix} \lambda_1 \\ \lambda_2 \end{matrix}\right\} = \lambda_n + F \pm F\sqrt{1+\lambda_n E} \tag{6.77}$$

with

$$D \equiv \frac{I'(0)}{I(\infty)-1}, \quad E \equiv \frac{I'(0)}{F^2}, \quad F \equiv D\frac{S(\infty)-3I(\infty)+2}{3[I(\infty)-1]} \tag{6.78}$$

where $I(\infty)$, $I'(0)$ and $S(\infty)$ are given in (6.69).

Suppose that the superposed packets are statistically multiplexed to transmit over a single line. If the packet length is generally distributed the system is modeled by MMPP/G/1, for which mean waiting time is calculated by (6.59).

[Example 6.7] Consider an ISDN in which the voice signal is packetized by 32 kb/s ADPCM (adaptive differential pulse code modulation).

With the sampling period $T = 16$ ms, the fixed packet length becomes

$$L = 32\,\text{kbps} \times 16\,\text{ms} = 512\,\text{bits} = 64\,\text{bytes}.$$

At transmission speed $v = 1.536$ Mb/s, the packet transmission (service) time is $h = L/v = (1/3)$ ms fixed, and hence we have $h^{(2)} = h^2$. Using the ON-OFF parameters $\alpha^{-1} = 352$ ms and $\beta^{-1} = 650$ ms [42], from (6.66) and (6.70) we have

$$\lambda = 0.02196/\text{ms}, \quad C_a^{\;2} = 18.0950, \quad S_k = 9.8379, \quad a(0) = 0.05819/\text{ms}.$$

For example, with the number of voice circuits $n = 120$, we have $\lambda_n = n\lambda = 2.6347$/ms, and from (6.77) the MMPP parameters,

$$r_1 = 0.002169, \quad r_2 = 0.002071, \quad \lambda_1 = 2.9509, \quad \lambda_2 = 3328/\text{ms}.$$

Hence, from (6.59), we have the actual mean waiting time $W_a = 3.5728$ ms.

The calculated result for various voice circuits is shown in Figure 6.5, which demonstrates good agreement with the simulation result.

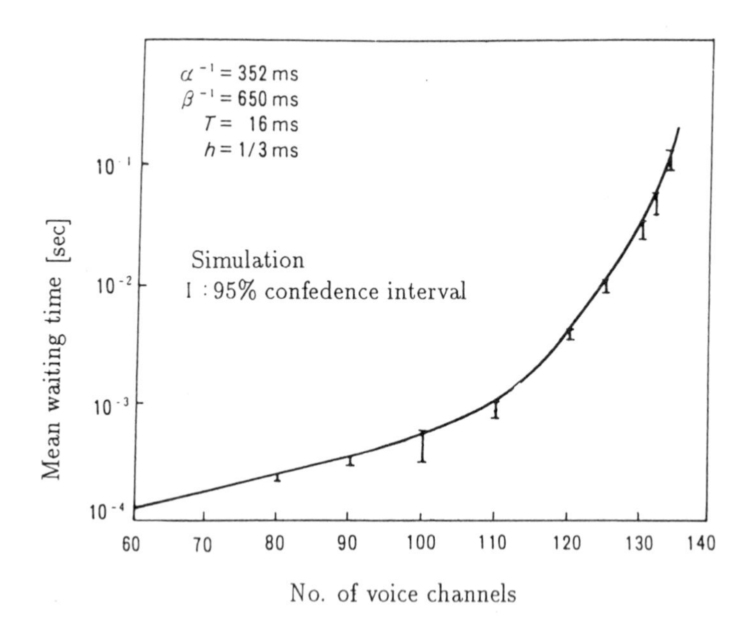

Fig.6.5 Mean waiting time for voice packet

6.4.4 MMPP + M/G/1

Consider the system integrating bursty (voice/video) and Poissonian (data) packet inputs to transmit over a single line as before. Approximating the bursty input by MMPP, the system is modeled by MMPP + M/G/1, and individual performances are evaluated as follows:

Denoting the MMPP parameters for the bursty input by r_1, r_2, λ_{b1} and λ_{b2}, from (6.61) the arrival rate λ_b for the bursty input is given by

$$\lambda_b = \frac{\lambda_{b1} r_2 + \lambda_{b2} r_1}{r_1 + r_2}. \tag{6.79}$$

Letting λ_p be the arrival rate for the Poisson input, the integrated process becomes again an MMPP with parameters, r_1, r_2, and

$$\lambda_1 = \lambda_{b1} + \lambda_p, \quad \lambda_2 = \lambda_{b2} + \lambda_p. \tag{6.80}$$

Thus, the system reduces to MMPP/G/1, and the individual mean waiting times W_b and W_p for bursty and Poisson inputs, respectively, are evaluated by using the PASTA.

Let W_i, $i = 1, 2$, be the mean virtual waiting times in phase i, and W_a and W_v be the mean actual and virtual waiting times, respectively, for the integrated traffic. Then, we have the relation,

$$\begin{aligned} &(\lambda_b + \lambda_p)W_a = \lambda_b W_b + \lambda_p W_p = \lambda_1 W_1 + \lambda_2 W_2 \\ &W_v = W_1 + W_2 = W_p \end{aligned} \tag{6.81}$$

where the last equality results from the PASTA for the Poisson input. Solving (6.81), we have the individual mean waiting times,

$$\begin{aligned} &W_b = \frac{\lambda_{b1} W_1 + \lambda_{b2} W_2}{\lambda_b} \\ &W_p = W_1 + W_2 \end{aligned} \tag{6.82}$$

where W_1 and W_2 are calculated by (6.65).

[Example 6.8] Consider the ATM system in which video and data are integrated, and packetized in cells of fixed 53-byte length to transmit over a single line with speed 156 Mb/s.

If the arrival processes of the video and data cells are approximated by MMPP and Poisson process, respectively, the system is modeled by MMPP + M/D/1. The cell transmission time is $h = 2.718\,\mu$s, and $h^{(2)} = h^2$ for the fixed cell. Approximate a single video source by the ON-OFF model with parameters $\alpha^{-1} = 2.208$ ms, $\beta^{-1} = 31.125$ ms, and a sampling period $T = 3.397\,\mu$s. For example, with 3 video sources, from (6.77) and (6.79) we get the MMPP parameters,

$$r_1 = 0.4145, \quad r_2 = 0.0705, \quad \lambda_{b1} = 365.65, \quad \lambda_{b2} = 6.24 \quad \lambda_b = 58.50/\text{ms}.$$

If the data cells arrive at Poisson rate $\lambda_p = 100\,/$ms, we have $\lambda_1 = 458.089\,/$ms and $\lambda_2 = 114.579\,/$ms. From (6.65) we get $W_1 = 0.1004$ ms and $W_2 = 0.0383$ ms. Using these in (6.82) yields the individual mean waiting times,

$$\text{Video}: \; W_b = 0.6316\,\text{ms}, \quad \text{Data}: \; W_p = 0.1387\,\text{ms}.$$

Figure 6.6 shows the calculated results for various data arrival rates. It should be noted that the delay-sensitive video cells suffer a longer delay than those for data, the latter being rather deferrable. This is due to the bursty nature of the video input, and some countermeasures such as priority control or buffering at the receiving end, are required in the design of the ATM system.

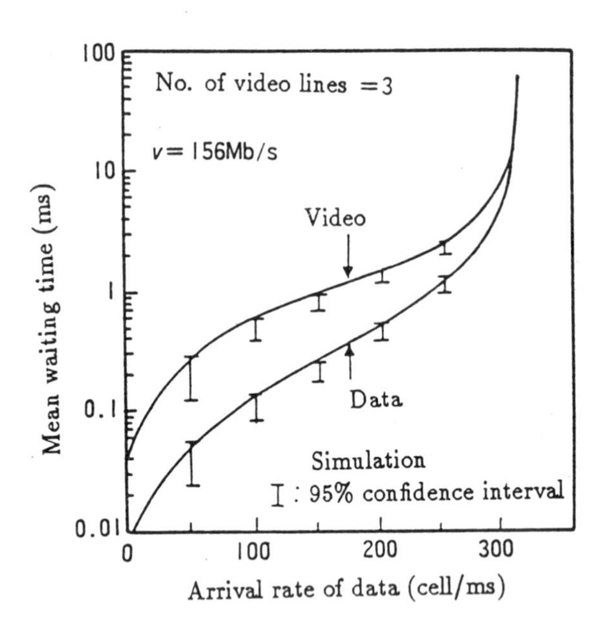

Fig.6.6 Mean waiting times for video and data cells

6.5 Local Area Networks

6.5.1 Token Ring LAN

The Token Ring LAN is a standard LAN protocol of sequential type access control. In this protocol, the transmission permission signal called a *token* circulates among a number of nodes allocated in a ring configuration. A node having data to be transmitted, seizes the free token and converts it to a busy token carrying the data packet with destination address. After the data packet has been received by the destination node, the token returns to the origination node to indicate so, where it is made free and again circulates in the ring. There are two types of making the token free:

(1) *Exhaustive type*: The token is made free after having transmitted all data including those received during the transmission.

(2) *Gated type*: The token is made free after having transmitted the data which had been waiting before the transmission began.

The Token Ring LAN is modeled as shown in Figure 6.7. The token is regarded as a single server, who walks around the ring to serve the nodes which have a call request (data to be transmitted). This is called the *walking server model*, and reduces to the cyclic multi-queue model described in Section 4.5.

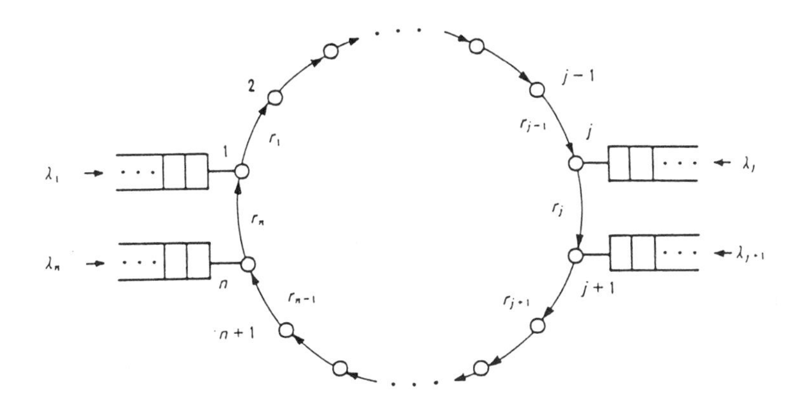

Fig.6.7 Token ring LAN

Let n be the number of nodes, and u_j and σ_{uj}^2 be the mean and variance of the *walking time* from node j to the next node, respectively. The walking time is the sum of the propagation time in the transmission line and processing time in the node. Denote the mean and second moment of the *cycle time* for node j by c_j and $c_j^{(2)}$, respectively. The cycle time is defined by the time interval from the server begins to serve the node until it returns to that node after rounding the ring.

For a symmetric system in which data arrive in Poisson process and the conditions of all nodes are identical, from (4.73) and (4.76) the mean waiting time for a node is given by, omitting the suffix $_j$,

$$W = \frac{\sigma_u^{(2)}}{2u} + \frac{(1 \pm \rho)c_0 + n\lambda h^{(2)}}{2(1-\rho_0)}. \tag{6.83}$$

The double sign $\pm$ takes $-$ for Exhaustive type, and $+$ for Gated type, λ is the Poisson arrival rate at a node, h and $h^{(2)}$ are the mean and second moment of the service time, respectively. $\rho = \lambda h$ is the node occupancy, $\rho_0 = n\rho$ the server occupancy, and $c_0 = nu$ the total walking time in the ring.

For an asymmetric system with the heterogeneous node condition, from (4.67) the mean cyclic time c_j is still independent of the node, and is given by

$$c_j = \frac{c_0}{1-\rho_0} \equiv c \tag{6.84}$$

where

$$c_0 = \sum_{j=1}^{n} u_j, \quad \rho_0 = \sum_{j=1}^{n} \rho_j.$$

Thus, from (4.69) and (4.74) the mean waiting time for node j is given by

$$W_j = (1 \pm \rho_j)\frac{c_j^{(2)}}{2c} \tag{6.85}$$

where $\pm$ has the same meaning as in (6.83), and the parameters for node j are indicated with suffix $_j$.

Equation (6.85) can be calculated by evaluating the second moment $c_j^{(2)}$ of the cycle time, but it needs a complex iteration with a large computer time [45]. A simple and fairly accurate approximation is proposed in [46]. This is described in the following subsection:

6.5.2 Approximation for Asymmetric System

The inter-visit time is defined as the time period from when the server leaves node j until it returns to that node. This is also referred to as *vacation time* for a walking server. Let v_j and $v_j^{(2)}$ be the mean and second moment of the *inter-visit time*, respectively. Then, from (4.68) we have the mean waiting time for node j,

$$W_j = \frac{v_j^{(2)}}{2v_j} + \frac{\lambda_j h_j^{(2)}}{2(1-\rho)}. \tag{6.86}$$

The right hand side is the sum of the mean residual vacation time and the M/G/1 mean waiting time in that node, and thus (6.86) holds for both Exhaustive and Gated types.

Averaging (6.86) weighted by ρ_j yields [45]

$$\sum_{j=1}^{n} \rho_j W_j = \frac{\rho_0}{1(1-\rho_0)} \sum_{j=1}^{n} \lambda_j h_j^{(2)} + \frac{c_0}{2(1-\rho_0)} \sum_{j=1}^{n} \rho_j (1 \pm \rho_j) \tag{6.87}$$

where $\pm$ has the same meaning as before, and a constant walking time with $\sigma_{uj}^2 = 0$ is assumed for simplicity. If $c_0 = 0$, (6.87) reduces to the M/G/1 conservation law in (4.29). Therefore, the second term in the right hand side represents the effect of the token ring scheme.

Recall from (6.84) that the mean cyclic time $c_j = c$ is independent of j. Therefore, we assume heuristically that the second moment $c_j^{(2)}$ is also independent of j. Thus, setting $c_j^{(2)} = c^{(2)}$ in (6.85), using it in (6.87) to obtain $c^{(2)}$, and substituting it into (6.85), we obtain the approximate formula,

$$W_j = \frac{1 \pm \rho_j}{2(1-\rho_0)} \left[c_0 + \frac{\rho_0 \sum_{j=1}^{n} \lambda_j h_j^{(2)}}{\sum_{j=1}^{n} \rho_j (1 \pm \rho_j)} \right] \tag{6.88}$$

where $\pm$ takes $-$ for Exhaustive type, and $+$ for Gated type as before.

[Example 6.9] Consider the asymmetric Token Ring LAN, in which $n = 20$, and nodes 1 and 8 are heavily loaded, where

$$\rho_j = 0.32,\ j = 1, 8;\quad \rho_j = 0.16/18 \doteq 0.0089,\ \text{others};\quad \rho_0 = 0.8$$
$$h_j = 1.072\,\text{sec},\ h_j^{(2)} = 2.196\,\text{sec}^2,\ j = 1, 2, \cdots, n;\quad c_0 = 0.025\,\text{sec}.$$

Table 6.2 shows the approximate results by (6.88) compared with the exact values in [45] [47]. It is seen that the approximation provides fairly good accuracy with a simple calculation.

Table 6.2 Mean Waiting Time for Token Ring LAN, in second

Node	ρ_j	Exhaustive		Gated	
		Approx.	Exact	Approx.	Exact
1	0.3200	3.796	3.798	4.382	4.370
2	0.0089	5.533	5.544	3.349	3.369
⋮	⋮	⋮	⋮	⋮	⋮
7	0.0089	5.533	5.596	3.349	3.351
8	0.3200	3.796	3.771	4.382	4.379
9	0.0089	5.533	5.504	3.349	3.382
⋮	⋮	⋮	⋮	⋮	⋮
20	0.0089	5.533	5.614	3.349	3.344

6.5.3 CSMA/CD LAN

The *CSMA/CD* (carrier sense multiple access with collision detection) is the standard LAN protocol of contention type access control. In this protocol, a node having a packet, transmits it after confirming the idle condition of the common bus by sensing no carrier on it. Due to the carrier propagation delay, however, there is the possibility of collision. If it happens the collided nodes re-transmit the packets after a predetermined time. This is called the *back-off* operation.

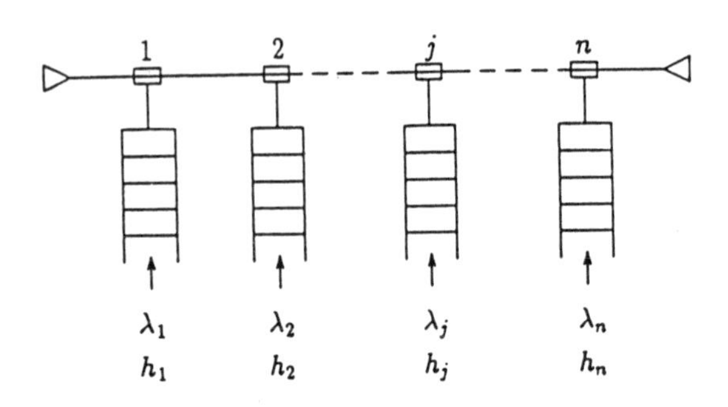

Fig.6.8 CSMA/CD LAN

A number of studies for the performance evaluation have been made on this protocol [48]. Here, somewhat simplified version of [49] is presented. The bus type CSMA/CD LAN is modeled as shown in Figure 6.8, in which n nodes are connected with a single bi-directional transmission bus. The following is assumed:

(1) In node j, $j = 1, 2, \cdots, n$, packets arrive at Poisson rate λ_j, for which transmission time h_j is fixed. For each node, an infinite buffer is provided with the FIFO discipline.

(2) Nodes are allocated with equal spacing, the propagation time between nodes i and j is τ_{ij}, and end-to-end propagation time $\tau = \tau_{1n}$.

(3) When the carrier terminates, each node is exhibited to access the bus for the *inter-frame gap*, which is exponentially distributed with mean t_{gap}.

(4) When a collision occurs, each node transmits a *jam signal* of length t_{jam} to ensure the back-off operation. The *re-transmission time* is exponentially distributed with mean μ^{-1}.

The mean waiting time W_j for node j is expressed by

$$W_j = W_{qj} + W_{aj} \tag{6.89}$$

where W_{aj} is the mean *access time* due to the CSMA/CD access control mechanism, and

$$W_{qj} = \frac{\rho'_j}{2(1-\rho'_j)}(1 + {C_{S_j}}^2)S_j \tag{6.90}$$

is the mean waiting time for an M/G/1 queue in the node, with

$$S_j = h_j + W_{aj}, \quad \rho'_j = \lambda_j(h_j + S_j) \tag{6.91}$$

and ${C_{S_j}}^2$ is the SCV corresponding to S_j.

The estimation of ${C_{S_j}}^2$ is somewhat involved [49], and here we approximate as ${C_{S_j}}^2 = 1$. Then, from (6.89) we obtain the approximate formula for individual mean waiting time,

$$W_j = \frac{\lambda_j {S_j}^2 + W_{aj}}{1 - \rho_j - \lambda_j W_{aj}} \tag{6.92}$$

where $\rho_j = \lambda_j h_j$. It remains to estimate the mean access time W_{aj}. This is described in the following.

6.5.4 Mean Access Time

The mean access time W_{aj} is expressed by

$$W_{aj} = p_{0j} t_{dj} + n_{cj} t_{bj} \tag{6.93}$$

where $p_{0j} = P\{$a packet arrives at node j in idle condition$\}$, t_{dj} is the mean delay due to the bus busy, n_{cj} is the mean number of collisions per unit time, and t_{bj} is the mean back-off time. These are calculated as follows [49]:

$$\begin{aligned}
p_{0j} &= 1 - \rho_j - \lambda_j W_{aj} \\
t_{dj} &= \sum_{k \neq j} \frac{\rho_k}{1 - \rho_j} \frac{h_k^{(2)}}{2h_k} \\
t_{bj} &= t_{jam} + \mu^{-1} + t_{dj} \\
n_{cj} &= \frac{p_{0j} p_{cj} + (1 - p_{0j}) p_{bj}}{1 - p_{cj}}
\end{aligned} \tag{6.94}$$

where $p_{cj} = P\{$collision occurs when node j transmits packet$\}$, and $p_{bj} = P\{$collision occurs in first transmission | packet arrives when node j is busy$\}$, are given by

$$\begin{aligned}
p_{cj} &= \sum_{k \neq j} \frac{\rho_k}{1 - \rho_j} P_k + \frac{1 - \rho_0}{1 - \rho_j} P_v \\
p_{bj} &= 1 - \prod_{i \neq j} [p_{0i} e^{-\lambda_i h_j} + (1 - p_{0i}) e^{-\mu h_j}]
\end{aligned} \tag{6.95}$$

with $\rho_0 = \sum_{j=1}^{n} \rho_j$. In (6.95), $P_k = P\{$collision occurs just after node j transmits packet$\}$ and $P_v = P\{$collision occurs when node j is idle$\}$, are given by

$$\begin{aligned}
P_k &= 1 - p_{0k} e^{-2\lambda_k \tau_{jk}} \prod_{i \neq j,k} [p_{0i} e^{-\lambda_i (h_k + \tau')} + (1 - p_{0i}) e^{-\mu (h_k + \tau')}] \\
P_v &= 1 - \prod_{i \neq j} [p_{0i} e^{-2\lambda_i \tau_{ij}} + (1 - p_{0i}) e^{-2\mu \tau_{ij}}]
\end{aligned} \tag{6.96}$$

where

$$\tau' = \begin{cases} 0, & i < j < k,\ k < j < i \\ 2\tau_{ij}, & j < i < k,\ k < i < j \\ 2\tau_{kj}, & j < k < i,\ i < k < j. \end{cases}$$

We can calculate the mean access time W_{aj} by iterating (6.93) to (6.96), and using (6.92) the individual mean waiting time W_j.

[Example 6.10] Consider a CSMA/CD LAN, known as Ethernet, in which coaxial cable is used for the bus, with transmission speed 10 Mb/s.

For example, let $n = 5$, $\tau = 5\,\mu$sec, $t_{jam} = 5\,\mu$sec, $t_{gap} = 10\,\mu$sec, and $\mu^{-1} = 50\,\mu$sec. If the packet size is fixed at 256 byte for all nodes, we have

$$h_j = 256 \times 8/10 = 204.8\mu\,\text{sec}, \quad h_j^{(2)} = h_j{}^2, \quad j = 1, 2, \cdots, 5.$$

With the arrival rate, $\lambda_j = 0.41$/ms, $j = 1, 3, 4, 5$ for light nodes, and $\lambda_2 = 0.81$/ms for heavy node, we have

$$\rho_j == 0.0840, \quad j = 1,3,4,5; \quad \rho_2 = 0.1679; \quad \rho = 0.5.$$

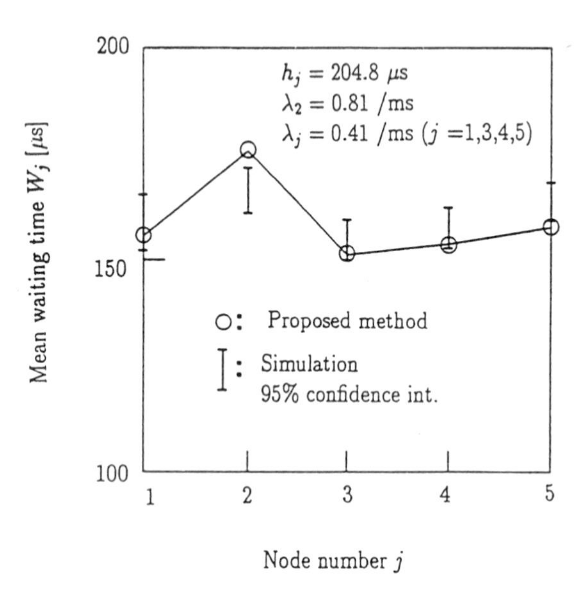

Fig.6.9 Mean waiting time for CSMA/CD LAN

The calculated result is shown in Figure 6.9. The approximation presents a fairly good agreement with the simulation result.

Exercises

[1] Suppose that a facsimile machine is located in each section in a company. Overflow documents from the sections are transmitted by 2 common-use facsimile machines in the administration section. With the arrival rate $\lambda = 90/\text{hr}$, and the SCV ${C_a}^2 = 2$ for the overflow process, obtain the mean waiting time for the common facsimile machine, using the modified diffusion approximation, under the assumptions:

(1) All the documents are of fixed form with transmission time 1 minute.

(2) Half of the documents are of the fixed form, and the transmission time of rest documents is exponentially distributed for the rest of the documents.

[2] In Example 6.2, show that the input is equivalent to an MMPP.

(1) Characterize the MMPP by the phase transition rate matrix R and arrival rate matrix Λ.

(2) Calculate for MMPP/M/$s(m)$ the blocking probability B and mean waiting time W with the parameters in Example 6.2 and $m = 1$.

[3] In packetized voice system described in Example 6.7, calculate the mean waiting time for the following parameters:

(1) $v = 1.536/2 = 0.768$ Mb/s, and $n = 60$.

(2) $v = 1.536 \times 2 = 3.072$ Mb/s, and $n = 240$.

[4] Consider 5 nodes uniformly allocated in a ring, with total transmission time $c_0 = 5\,\mu\text{sec}$. With the same traffic condition as in Example 6.10, calculate the mean waiting times of each node,

(1) For an Exhaustive-type Token Ring LAN.

(2) For a Gated-type Token Ring LAN.

Chapter 7

TRAFFIC SIMULATION

Since the telephone was invented, a number of teletraffic models have been studied as discussed in preceding chapters. However, there are many models for which exact solutions cannot be obtained analytically. This chapter explains the use of computer simulations to solve such traffic problems.

7.1 Introduction

Exact analytical solutions in teletraffic theory are sometimes limited in capability of performance analysis of large-scale communications networks, switching control systems with complex scheduling strategies, congestion control and routing control, etc. Computer simulation has therefore been applied as an alternative approach. Table 7.1 compares simulation with analytical solution and numerical analysis.

Table 7.1 Solutions for teletraffic problems

Solution	Application ranges	Accuracy of result	Computation time	Memory required	Program size
Analytical method	Narrow	High	Short	Small	Small
Numerical analysis	Medium	Medium	Medium	Medium	Medium
Simulation	Wide	Low	Long	Large	Large

The simulation is useful not only for the case that the analytical solution is not obtained, but also in the cases where numerical computation of an analytical solution is difficult, and for validating the accuracy of approximate solutions. Simulation is also used in various stages in design and operation of teletraffic systems such as:

(1) Clarifying teletraffic problems to feedback to the system design.

(2) Evaluating the performance to confirme the system application range.

(3) Estimating the response to fault and over-load to determine the countermeasures.

For Item (1), it is important to get the result quickly and timely for the system design, rather than to pursue a higher accuracy. Therefore, limited-scale simulations are often used, which can be quickly built and changed easily by using a simulation language. On the other hand, for Items (2) and (3), full-scale simulations are required to obtain precise traffic characteristics for dimensioning the system.

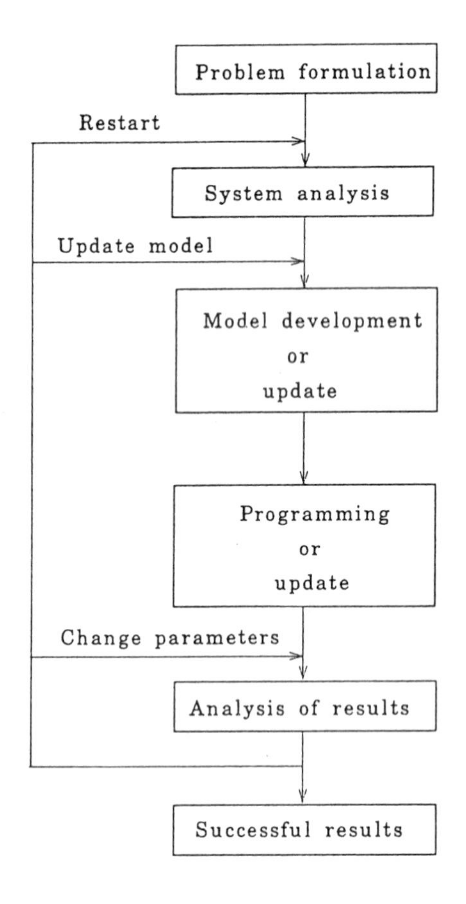

Fig.7.1 Process of simulation

Figure 7.1 shows the process of simulation from planning to solution of the problem. In the first step, the problem is formulated, and the performance measures of the system are specified. For example, in control system for electronic switching, the connecting time and call processing capacity are the performance measures. Once the problem has been described, the system analysis follows in order to obtain information for modeling the system. Based on this information, a model is established to solve the problem. The process up to this stage is the most important.

After establishing the model, it remains only to convert it to the program for performing the simulation. For this process, it is necessary to make a reasonable plan for memory allocation for data and programs, and computation time, according to the performance of the computer. In executing the simulation, consideration should be given to run time and the number of operations, particularly if the budget for computation is limited.

When the simulation result has been obtained, it is used for performance evaluation and analysis of the system in question. In this process, it is most important to determine whether the intended simulation has been performed and a valid result obtained. The data obtained are used to analyze and estimate the system for achieving the optimum design.

Methods for simulation in accordance with the above procedures, are described in the following sections.

7.2 Methods of Simulation

7.2.1 Problem Formulation and Modeling

The formulation of the problem and its modeling are placed in the beginning of the simulation procedure, and they are important for achieving an efficient simulation. The model in the simulation should abstract the essence of the system in question, and be as simple as possible to facilitate the simulation. It is difficult to estimate theoretically the validity of a model, so the expertise of persons making the models has to be relied upon. Here, we shall describe on the modeling world-view [50].

In general, there are two kinds of simulations, continuous and discrete, and a traffic simulation is classified as a discrete-event simulation. That is, the discrete state (number of calls) of the system is changed by an event of call origination or termination. As shown in Figure 7.2, there are three viewpoints for modeling in a discrete-event simulation: *event, process* and *activity.*

(1) Event-oriented Modeling

A simulation model is formulated by describing state changes by events such as call origination and termination. This is used in the cases where general purpose programming languages, such as FORTRAN and PL/1, are used.

(2) Process-oriented Modeling

This modeling describes the behavior of the entity (call) in the system. This is also called *entity-oriented modeling*, and used in cases where simulation languages, such as GPSS and SLAM II, are used for describing simulation models.

(3) Activity-oriented Modeling

This modeling describes the time instants of the initiation and termination of activities such as call duration and trunk busy. This method is suitable for modeling systems for which holding time depends on the state of the system (number of existing calls). However, since the simulation is performed by scanning the activities, the execution time is longer than for event-oriented modeling.

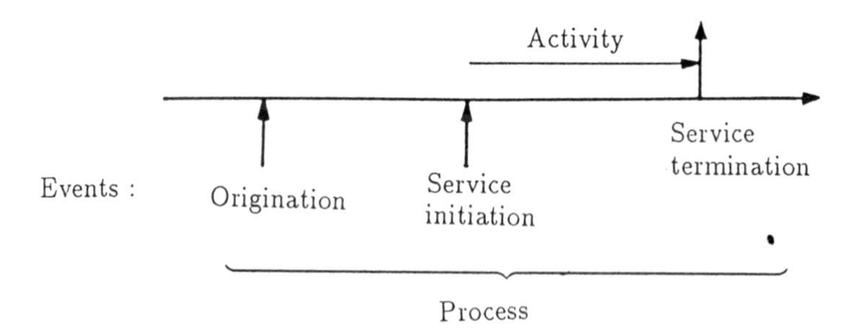

Fig.7.2 Viewpoints in simulation

7.2.2 Programming for Simulation

There are two kinds of programming languages for simulation: *simulation language* and *general purpose language*. The former includes GPSS, SIMSCRIPT and SLAM II, and the latter includes FORTRAN, PL/1 and ALGOL. Typical simulation languages are shown in Table 7.2.

Simulation languages include the following functions:

(1) Generation of random numbers.

(2) Execution of time schedules.

(3) Data saving, statistical analysis, output format conversion.

The advantages of using simulation languages are the elimination of the need to program the functions listed above, and the ease of programming once the process oriented model is formulated.

Table 7.2 Typical simulation languages

Language	GPSS	SIMSCRIPT	SLAM II
Prerequisite knowledge	None	FORTRAN	None (interface to FORTRAN)
Symbols for flow chart	Special	Not special	Special
Modeling method	Process-oriented	Event-oriented	Process-oriented Event-oriented
Programming unit	Block	SIMSCRIPT sentence Event routine	Node Activity
Simulation entity	Transaction	Temporary entity	Entity

On the other hand, general purpose languages are more flexible than simulation languages. Moreover, they can achieve a faster execution time. Choice of languages depends on factors such as convenience of usage, knowledge of programmer, characteristics of models, etc. In order to facilitate learning of the languages, simulation support systems have been developed, for example a simulator with graphical input [51].

7.2.3 Simulation Language, GPSS

The most popular simulation language is GPSS (general purpose simulation system), and its outline is given here. The GPSS describes a simulation program by process-oriented modeling, using about 40 block diagrams written in special symbols. An entity moving according to the diagrams is called a *transaction*, and it is chained basically in one of the two chains:

(a) *Current event chain*: Transactions, whose scheduled moving times are before or at the present clock time, are chained according to an order of the priority, or scheduled moving times. They have not yet been moved due to some conditions not satisfied, and are kept waiting for moving.

(b) *Future event chain*: Transactions, whose scheduled moving times are after the present clock time, are chained in order according to moving time.

Having completed all processes at the present clock time, the system program sets the clock to the scheduled moving time, say t_1, for the transaction at the top of the future event chain. Then, all the transactions in the chain having the moving time equal to t_1, are transferred to the current event chain, where the order of transactions are rearranged including previously chained transactions. The top transaction in this newly arranged chain, proceeds to the predetermined block. The procedure continues until the transaction cannot move due to the condition not satisfied, or it enters a block for which sojourn time is not zero. When entering a block with sojourn time $t_s > 0$, the transaction is transferred to the future event chain, with the scheduled moving time $t_1 + t_s$. If the states of the storages and facilities are changed as a result of a transaction moving, the system program returns to the top transaction in the current event chain to examine if it is movable, and if so it is moved. When no transactions in the current event chain are movable, the system program returns to the previous point, and sets the clock to the time for the next process to be made.

Although there are additional chains other than (a) and (b), the basic control mechanism of GPSS is as stated above. The advantages of GPSS are:

(1) The language is simple and easy to understand.

(2) Statistics, such as the mean and variance of queue length or waiting time, are automatically calculated and displayed in histograms.

(3) Formulation and modification of models are easy.

The disadvantages of GPSS are:

(1) Only the system functions SAVEVALUE and VARIABLE are defined arbitrarily, but the standard functions, such as exponential function, cannot be used.

(2) Long computer time is required for execution.

(3) Arbitrary printing format cannot be used, since the format is specified.

7.2.4 General Purpose Programming Languages

There are two execution methods, the *Markov chain method* and the *time tracing method*, for simulations using general purpose programming languages.

(1) Markov Chain Method

This is also referred to as the *roulette model*, and directly simulates the Markov chain representing the system state changes, regardless of the time progress. The principle is shown in Figure 7.3. It is based on the fact that the system state (number of calls) increases or decreases by one according to call arrival or termination, respectively. Probabilities of call arrival and termination are determined by the arrival rate or service rate.

Although the Markov chain method is applicable only to systems for which state changes are described by a Markov chain, it requires less computation time and memory for simulation compared with the time tracing method. Therefore, the method is efficient for simulation of non-delay systems appearing in speech path networks.

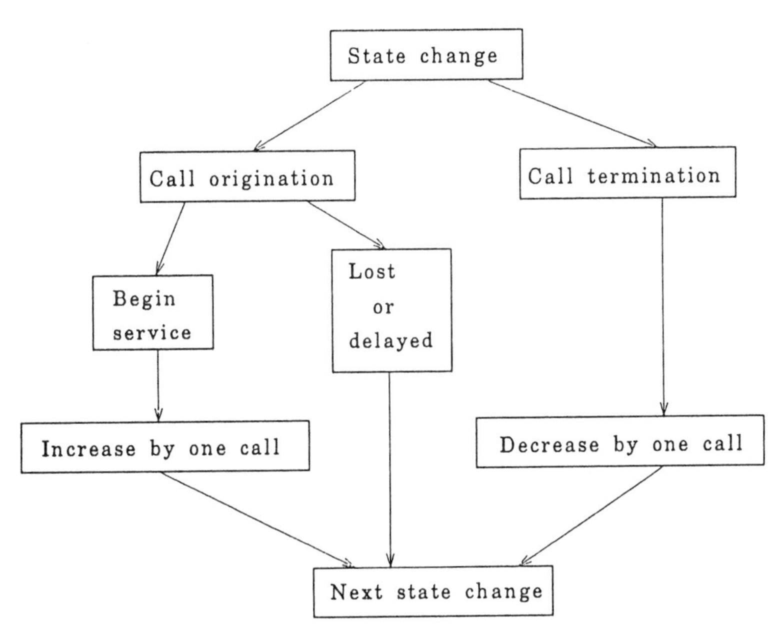

Fig.7.3 Principle of Markov chain method

[Example 7.1] The Markov chain method is explained for an M/M/$s(0)$ loss system.

Assume Poisson traffic load a erl, and exponential service time with mean h. Then, the probability that a call originates in an infinitesimal interval Δt, is $(a/h)\Delta t$. Letting n be the number of calls present in the system, the probability of a call terminating in Δt is $(n/h)\Delta t$. When a call originates or terminates, the system state changes as $n \to (n+1)$ or $n \to (n-1)$, respectively, for $1 \le n \le s-1$. For $n = 0$, no termination occurs, and for $n = s$, an arriving call is lost, which is also regarded as a state change.

Given that a state change occurs, the conditional probability that the change is a call arrival, is given by, from the Baye's theorem (See Appendix C [T2].)

$$p_a = \frac{(a/h)\Delta t}{(a/h)\Delta t + (n/h)\Delta t} = \frac{a}{a+n}. \tag{7.1}$$

Likewise, the conditional probability that the change is a call termination is given by

$$p_b = \frac{n}{a+n}. \tag{7.2}$$

Generate the random variable Y following the uniform distribution $U(0,1)$, that is

$$P\{Y \leq y\} = \begin{cases} 1, & y \geq 1 \\ y, & 0 < y < 1 \\ 0, & y \leq 0. \end{cases} \tag{7.3}$$

Then, starting from a certain state, the state is changed

$$\begin{aligned} &\text{If } 0 < Y < a/(a+n), \text{ then a call arrives, and } n \to n+1, \text{ or} \\ &\text{If } a/(a+n) < Y < 1, \text{ then a call terminates, and } n \to n-1. \end{aligned} \tag{7.4}$$

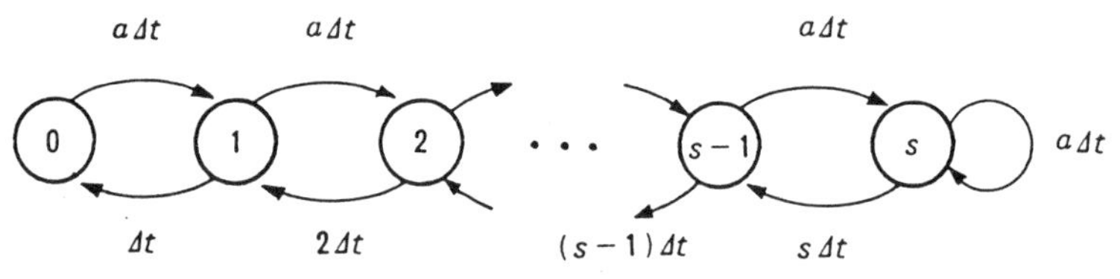

ⓝ : The number of calls in the system.

$a\Delta t$: Call origination probability in Δt.

$n\Delta t$: Call termination probability in Δt with $h = 1$.

Returns in state ⓢ with probability $a\Delta t$ indicate lost calls.

Fig.7.4 State transition in Markov chain method

The process of state changes is shown in Figure 7.4. A right-arrow path indicated with $a\Delta t$, corresponds to an arriving call, and a return path at state s corresponds to a lost call. The blocking probability is calculated by the ratio of lost calls to the total arriving calls. The distribution function and mean number of calls present in the system are also obtained by counting that number at the call arrival epoch, from the PASTA for Poisson arrivals.

(2) Time Trace Method

This method traces the time epochs at which call arrival or termination occurs, and the principle is shown in Figure 7.5. The programming environment provides an *event calendar*, a table listing the events to occur in the future. Usually, the arrival time of the first call is written in the event calendar as the initial value. During the progress of the simulation, time epochs of the next arrival call and termination of calls in service, are determined by random number, and written in the calendar. According to the calendar, the program proceeds the clock time to that of the nearest event, and repeats the process to execute the simulation.

The time trace method can specify an arbitrary distribution for interarrival time and service time, and hence it is applicable to the performance evaluation of systems involving complex scheduling strategies.

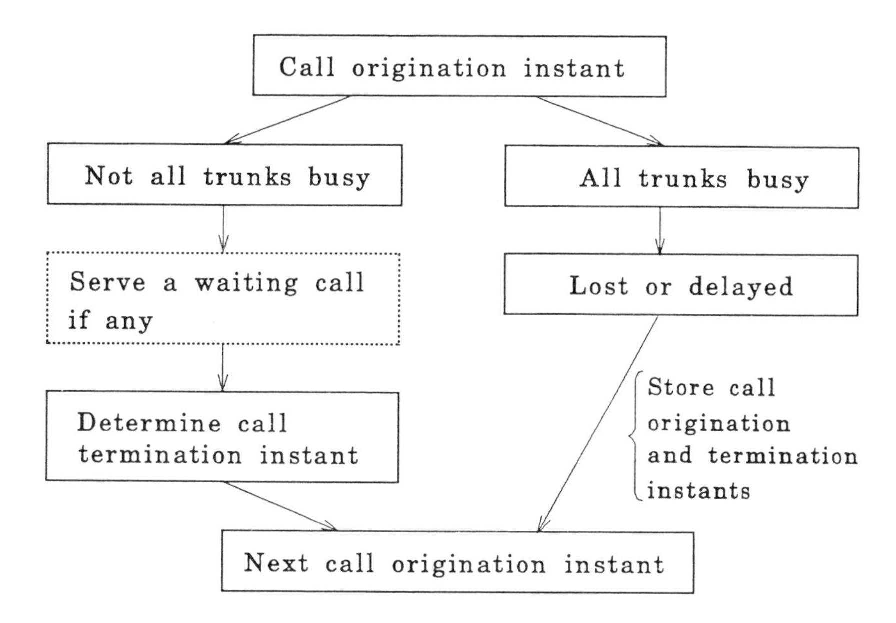

Fig.7.5 Principle of time trace method

7.3 Generation of Random Numbers

7.3.1 Random Number Generation Methods

There are two kinds of random numbers: *arithmetic random numbers* generated by a computer, and *physical random numbers* generated by phenomena such as radio-active substances or cosmic rays. The arithmetic random number is a series of non-random numbers in a strict sense, with a sufficiently long repeating period. Thus, they are called *pseudo-random numbers*, and the randomness is to be checked.

The physical random number is generated, for example, by using the gamma ray radiated following a Poisson distribution, and a complete randomness is obtained in principle. Due to the difficulty of generating identical random numbers, however, this is not suitable for traffic simulations which require repeated executions.

The required condition for pseudo-random numbers is:

(1) Sufficiently many random numbers can be readily generated.

(2) A sufficient randomness is obtained with a long repeating period.

(3) Statistically identical random numbers can be generated repeatedly.

(4) The statistical characteristic meets the simulation objectives.

7.3.2 Uniform Random Number

Random number distributed uniformly in a range are called *uniform random numbers.* These are also used for generating random numbers with an arbitrary distribution. Typical methods of generating the uniform random number are described.

(1) Mid-Square Method

This is the first generation method invented by von Neuman. Square the number x_0 with $2n$ figures, take the $2n$-figure number in the middle, and let it be x_1. Repeat similar operations to make a series of $x_0, x_1, x_2, \cdots$. For example, with $n = 2$,

$$x_0 = 2061, \qquad x_1 = 2477, \qquad x_2 = 1355$$

$$x_0{}^2 = 04247721, \quad x_1{}^2 = 06135529.$$

Since this method has a relatively short repeating period, it is no longer used.

(2) Multiplicative Congruential Method

A series of random numbers, $x_0, x_1, x_2, \cdots$, is obtained by the computation,

$$x_n = kx_{n-1} \pmod{M} \tag{7.5}$$

where $(\bmod M)$ represents the modulo M, the residue of division by M. It is known that the maximum repeating period 2^{b-2} is obtained for $k = \pm 3 \pmod 8$, $x_0 = 1 \pmod 2$ and $M = 2^b$.

(3) Mixed Congruential Method

A series of random numbers, $x_0, x_1, x_2, \cdots$, is obtained by the computation,

$$x_n = (kx_{n-1} + \mu) \pmod{M}. \tag{7.6}$$

The maximum period 2^b is obtained for $k = \pm 3 \pmod 8$, $x_0 = 1 \pmod 2$ and $M = 2^b$.

(4) M Sequence Method

A sequence of random numbers, $x_0, x_1, x_2, \cdots$, is obtained by the computation,

$$x_n = \sum_{i=1}^{k} c_i x_{n-1} \pmod 2 \tag{7.7}$$

with the maximum period $2^k - 1$, where $c_1, c_2, \cdots, c_{k-1} = 0$ or 1, and $c_k = 1$. Making binary decimals by consecutive l $(\leq k)$ numbers taking from the series $\{x_j\}$, constitutes a random number uniformly distributed in (0,1).

This method requires a number of repeated operations for obtaining a random number, since only one binary random number can be generated per computation. Although it seems disadvantageous in the point of speed, but since the mod 2 operation is equivalent to that of exclusive-or logic, the generation of random number may be speeded up by parallel execution of the logical operation.

7.3.3 Arbitrary Random Numbers

Generation methods are described here for random numbers following arbitrary distributions, which are often required in simulation. Unfortunately, however, no general method has been developed, which gives an algorithm providing program for a given distribution function, except for approximate methods. Therefore, it is necessary at present to select a suitable algorithm from several methods according to the distribution function to be implemented. In traffic simulations, algorithms for generating random number have to be not only exact but also high speed. Therefore, knowledge is needed for computers as well as statistics, when developing a simulator.

Typical methods for generating arbitrary random numbers are presented below.

(1) Inverse Transform Method

This method provides an algorithm for generating random numbers following a distribution function $F(\cdot)$ for which an inverse function $F^{-1}(\cdot)$ exists. The principle of this method is shown in Figure 7.6.

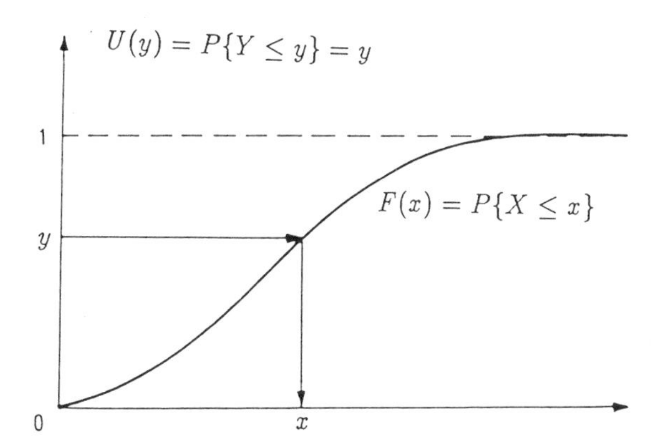

Fig.7.6 Principle of inverse transform method

Letting Y be a random variable following the uniform distribution in (0,1), we have

$$U(y) = P\{Y \leq y\} = y, \quad 0 < y < 1. \tag{7.8}$$

Generate the uniform random number Y by a method described before, and let

$$X = F^{-1}(Y). \tag{7.9}$$

Then, X follows the distribution function $F(x)$. This is verified from the relation,

$$P\{X \leq x\} = P\{F^{-1}(y) \leq x\} = P\{Y \leq F(x)\} = F(x). \tag{7.10}$$

[Example 7.2] Suppose that we want to have the random number X following the exponential distribution,

$$F(x) = 1 - e^{-\lambda x}, \quad x \geq 0. \tag{7.11}$$

From (7.9), we have

$$X = F^{-1}(Y) = -\frac{\log(1-Y)}{\lambda}. \tag{7.12}$$

It is worth noting here that if Y follows the uniform distribution of (7.8), from the relation,

$$P\{(1-Y) \leq y\} = P\{Y \geq 1-y\} = 1 - U(1-y) = y \tag{7.13}$$

$(1-Y)$ also follows the uniform distribution. Therefore, the equivalent result is obtained by the relation,

$$X = F^{-1}(1-Y) = -\frac{\log Y}{\lambda}. \tag{7.14}$$

In fact, this is verified from

$$\begin{aligned} P\{X \leq x\} &= P\left\{-\frac{\log Y}{\lambda} \leq x\right\} = P\{Y \geq e^{-\lambda x}\} \\ &= 1 - U(e^{-\lambda x}) = 1 - e^{-\lambda x}. \end{aligned} \tag{7.15}$$

For generating the random number following a discrete distribution with $F(x)$ of a step function, a similar method is applied using the accumulating probability table, which provides the inverse function. (See Exercise[1](2) in this chapter.)

(2) Rejection Method

Consider a distribution for which density function $f(x)$ is bounded. Let A be the area between $y = f(x)$ and the x axis. Then, the coordinate x of a random sample taken from the area A becomes a random number following $f(x)$.

Let $g(x)$ be a function which is larger than or equal to $f(x)$, and suppose that the random variable following

$$\overline{g}(x) = C^{-1}g(x) \tag{7.16}$$

is generated, where the integral,

$$C = \int g(x)ds \tag{7.17}$$

is bounded.

The random number following $f(x)$ is generated by the following algorithm:

[Algorithm 7.1]

(1) Generate the random numbers X and Y following $\overline{g}(x)$ and $U(0,1)$, respectively.

(2) If $Y > f(X)/g(X)$, then return to (1). Otherwise, X is the random number required.

The name of this method comes from the fact that X is rejected until a point belonging to the area A is sampled. The efficiency of this method for generating random numbers depends on:

(a) The mean number $(C-1)$ of rejections to get a random number.

(b) The speed generating the random number following $\overline{g}(x)$.

(c) The computing process of $f(X)/g(X)$.

In order to improve the efficiency, the selection of a suitable function $g(\cdot)$ is most important, and Item (c) has also a significant effect. If the density function is not bounded, a suitable bounded density function is found by transformation of the variable, and then the rejection method may be applied to the bounded function.

(3) Composition Method

Suppose that a probability density function $f(x)$ is expressed in integral form, such as

$$f(x) = \int f(x,\theta)g(\theta)d\theta \tag{7.18}$$

in terms of functions $f(x,\theta)$ and $g(\theta)$.

If we generate first the random number θ_0 following $g(\theta)$, and next the random number x_0 following $f(x,\theta)$, it is clear that x_0 follows $f(x)$. This is called the *composite method*, and in practice, if $g(\theta)$ is discrete,

$$f(x) = \sum r_k f_k(x) \tag{7.19}$$

is often used.

If we can use a relatively large value of r_k corresponding to $f_k(x)$ following which a random number is readily generated (e.g. the uniform distribution), the random number required may be generated efficiently.

7.4 Analysis of Output Result

7.4.1 Confidence Interval

Traffic simulation is interpreted as "the application of random sampling methods to deterministic and probabilistic problems". Therefore, the estimation with *confidence interval* is often applied for sample mean $\overline{X}$ from the simulation result. The shorter the interval, the higher is the estimation.

The confidence limit is calculated by grouping the simulation results into appropriate batches. The methods include the *replication method*, *batch mean method* and *regenerative method*. Figure 7.7 illustrates the concepts of these methods.

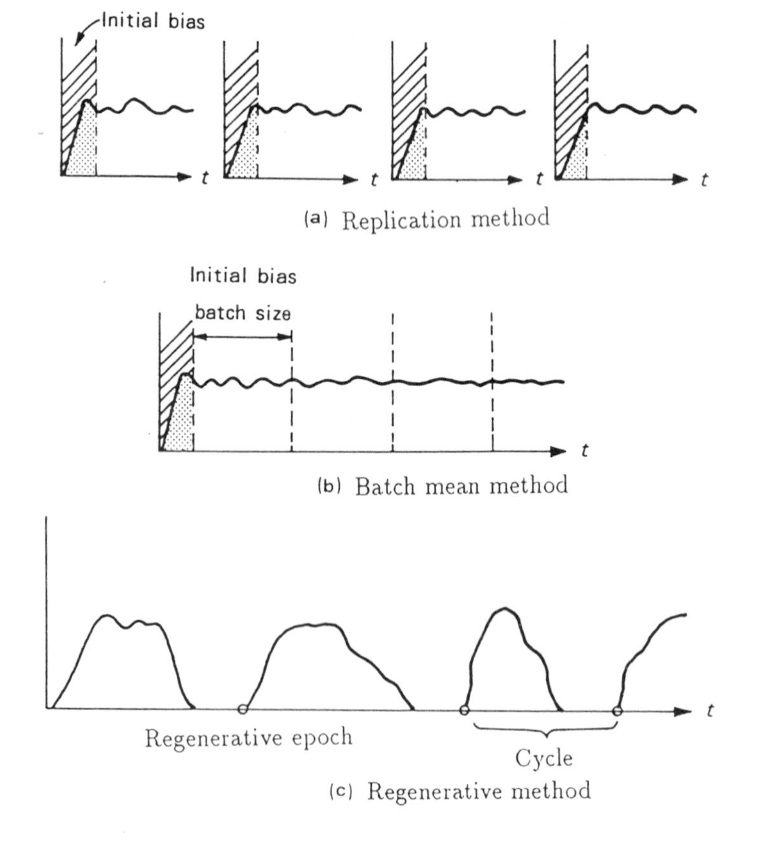

Fig.7.7 Calculation methods for confidence interval

7.4.2 Replication Method

A number of independent simulations are made, and the mean and confidence interval are calculated, by rejecting results in the initial transient stage according to:

(1) Simulation is started from a certain initial state, using a priori information, for example on equilibrium state, or

(2) Data is rejected before a predetermined time T, starting from the empty initial state.

Although a provisional test is desirable for determining T, it is usually determined by inspection or heuristically, and the process of rejecting the initial data is called *rejection of initial bias.*

If observed values $X_1, X_2, \cdots, X_n$ for a value X, say the number of calls in the system, are independent, the confidence interval $[\mu_\alpha{}^+, \mu_\alpha{}^-]$ of the mean $\mu = E\{X\}$ with probability $(1-\alpha)$ is calculated as follows:

Letting σ^2 be the variance of X, from the *central limit theorem* (See Appendix C.4 [T7].), the sample mean,

$$\overline{X} = \frac{1}{N}\sum_{i=1}^{n} X_i \tag{7.20}$$

approximately follows $N(\mu, \sigma^2/n)$, normal distribution with mean μ and variance σ^2/n, as $n \to \infty$. Using this theorem, the *confidence limits* $\mu_\alpha{}^+$ and $\mu_\alpha{}^-$ with probability $(1-\alpha) \times 100\,\%$ are determined by

$$\left.\begin{matrix} \mu_\alpha{}^+ \\ \mu_\alpha{}^- \end{matrix}\right\} = \overline{X} \pm \frac{\sigma}{\sqrt{n}} u_{\alpha/2} \tag{7.21}$$

where $u_{\alpha/2}$ is the value of the normal distribution such that

$$P\{X > u_{\alpha/2}\} = \alpha/2. \tag{7.22}$$

Usually σ^2 is unknown, and therefore from the simulation result the *sample variance,*

$$S^2 = \frac{1}{n-1}\sum_{i=1}^{n}(X_i - \overline{X})^2. \tag{7.23}$$

is calculated which is the non-biased estimate for σ^2. (See Appendix C [T4].) In this case, instead of $u_{\alpha/2}$, the corresponding value $t_{\alpha/2}{}^{n-1}$ of the *t distribution* with degree of freedom $n-1$ is used. (See Appendix C.4.3[T7].) Table 7.5 shows the value of $t_{\alpha/2}{}^{n-1}$ for probability $1-\alpha = 95\,\%$.

Table 7.5 Value of t distribution for 95 % confidence interval

No. of samples (n)	Degree of freedom ($n-1$)	${t_{0.05/2}}^{n-1}$
2	1	12.706
3	2	4.303
4	3	3.182
5	4	2.776
6	5	2.571
7	6	2.447
8	7	2.365
9	8	2.306
10	9	2.262
11	10	2.228
12	11	2.201
13	12	2.179

7.4.3 Batch Mean Method

In this method, n samples obtained by a simulation run are partitioned into M batches with $n' = n/M$ samples each. Then, the mean and the confidence interval are calculated from those of M batches. Letting X_{mi} be the ith samples in the mth batch, the mean in the batch is given by

$$\overline{X}_m = \frac{1}{n'} \sum_{i=1}^{n'} X_{mi} \tag{7.24}$$

and the overall sample mean by

$$\overline{X} = \frac{1}{M} \sum_{m=1}^{M} X_m. \tag{7.25}$$

The number M of batches should be as large as the central limit theorem holds. Although it seems desirable that some data between batches be rejected to remove the correlation between batches, it is known that the variance of samples may be reduced rather by utilizing these data. Therefore, in most cases after rejecting the data during the initial period T as before, all data are used for the batch mean method in most cases.

In practice, it is known that with $n' > 25$ the correlation between batches becomes negligible, and the batch mean $\overline{X}_m$ approximately follows the normal distribution $N(\mu, \sigma^2/n')$, where μ and σ^2 are the mean and variance of X, respectively. Hence, similarly to (7.21), the confidence limits are calculated by

$$\left.\begin{array}{l}\mu_\alpha^+\\ \mu_\alpha^-\end{array}\right\} = \overline{X} \pm \frac{\sigma_m}{\sqrt{M}} u_{\alpha/2} \tag{7.26}$$

where σ_m^2 is the variance of $\overline{X}_m$. If σ_m^2 is unknown, and instead

$$S^2 = \frac{1}{M-1} \sum_{m=1}^{M} (X_m - \overline{X})^2 \tag{7.27}$$

is used, the value of ${t_{\alpha/2}}^{M-1}$ in Table 7.5 is also used instead of $u_{\alpha/2}$ as before.

7.4.4 Regenerative Method

A stochastic process $\{X(t); t \geq 0\}$ is called a *regenerative process*, if it behaves with a probabilistic rule independent of past history, at time series $\{t_n\}$ which is referred to as the *regenerative epoch*. For example, in a delay system, letting $\{t_n\}$ be call arrival epochs when the system is empty, the number of calls existing in the system constitutes a regenerative process with regenerative epochs $\{t_n\}$.

The *regenerative method* is based on the principle that the batches are formed between regenerative epochs, and has advantages in stochastic treatment. For example, we can determine in advance the length of simulation computer time needed to obtain a predetermined accuracy in the simulation result. However, there remain some problems with applying this method to practical models, for example how to determine the regenerative epochs. For further details, refer to [4, p.290].

Exercises

[1] Using the uniform random numbers distributed in $(0,1)$, generate

(1) The random numbers with the density function,

$$f(x) = \begin{cases} 2x, & 0 \leq x \leq 1 \\ 0, & \text{otherwise.} \end{cases}$$

(2) The random numbers following the Poisson distribution with mean 3.

[2] From the simulation of a traffic model, the mean queue lengths in 10 batches are obtained as shown in the table below.

(1) By applying the batch mean method, estimate the mean queue length and the 95 % confidence interval.

(2) If the arrival rate is $\lambda = 0.5$/sec and the mean service time is $h = 1$ sec, estimate the mean waiting time.

Batch No.	Mean queue length
1	1.6066
2	1.1960
3	1.6180
4	2.0105
5	0.9609
6	1.4882
7	1.5795
8	2.2253
9	1.7656
10	1.4930

Bibliography

[1] Akimaru, H., Kawashima, K., Teletraffic - Theory and Applications, Telecommunications Association, 1990. (in Japanese)

[2] Syski, R., Introduction to Congestion Theory in Telephone Systems, Oliver & Boyd, 1960.

[3] Fujiki, M., Gambe, E., Teletraffic Theory, Maruzen, 1980. (in Japanese)

[4] Cooper, R.B., Introduction to Queueing Theory, Second Ed., North Holland, 1981.

[5] Akimaru, H., Cooper, R.B., Teletraffic Engineering, Ohm Pub. Co., 1985. (in Japanese)

[6] Akimaru, H., Ikeda, H., Switching Systems Engineering, Ohm Pub. Co., 1989. (in Japanese)

[7] Schwartz, M., Telecommunication Networks, Wisley, 1987.

[8] Kleinrock, L., Queueing Systems, Vol.I, II, John Wiley & Sons, 1975.

[9] Gelenbe, E., Mitrani, I., Analysis and Synthesis of Computer Systems, Academic Press, 1980.

[10] Çinlar, E., Introduction to Stochastic Processes, Prentice-Hall, 1975.

[11] Walff, R.W., Poisson Arrivals See Time Averages, Oper. Res., Vol.30, pp.223-231, 1982.

[12] Melamed, B., Whitt, W., On Arrivals that See Time Averages, Operations Research, Vol.38, No.1, 1990.

[13] Crommelin, C.D., Delay Probability Formula, POEEJ, Vol.26, pp.266-274, 1933.

[14] Okuda, T., Akimaru, H., Nakao, F., Analysis of Modified Erlangean Input Single Server Model, $\hat{E}_k/G/1$, IEEE Trans. COM., to appear, 1992.

[15] Kraemer, W., Langenbach-Belz, M., Approximate Formula for General Single Server and Batch Arrivals, Angew. Inf., Vol.20, 1978.

[16] Jaiswal, N.K., Priority Queues, Academic Press, 1968.

[17] Gimpel, L.A., Analysis of Mixture of Wide and Narrow Band Traffic, IEEE Trans. on Communication Tech., Vol.13, No.3, 1965.

[18] Kawashima, K., A Numerical Analysis for Delay-Delay Trunk Reservation System, Trans. of IEICE, Vol.62-B, No.7, 1979. (in Japanese)

[19] Cohen, J.W., Certain Delay Problems for a Full Availability Trunk Group Loaded by Two Traffic Sources, Communications News, Vol.16, No.3, 1956.

[20] Bhat, U.N., Fischer, M.J., Multichannel Queueing Systems with Heterogeneous Classes of Arrivals, Naval Research Logistics Quarterly, Vol.233, No.2, 1976.

[21] Akimaru, H., Kuribayashi, H., Katayama, A., Approximate Evaluation for Mixed Delay and Loss Systems with Renewal and Poisson Inputs, IEEE Trans. COM-36, No.7, 1988.

[22] Kuehn, P.J., Approximate Analysis of General Queueing Networks by Decomposition, IEEE Trans. COM-27, No.1, 1979.

[23] Takagi, H., Analysis of Polling Sytems, The MIT Press, 1988.

[24] Hashida, O., Analysis of Multi-Queues, Review of ECL, Vol.20, Nos. 3-4, 1972.

[25] Heffes, H., Analysis of First-Come-First-Served Queueing System with Peaked Inputs, BSTJ, Vol.52, pp.1215-1228, 1973.

[26] Descloux, A., On Overflow Process of Trunk Groups with Poisson Inputs and Exponential Service Times, BSTJ, Vol.42, pp.383-397, 1963.

[27] Niu, Z.S., Akimaru, H., Analysis of Overflow Traffic from Mixed Delay and Non-delay Systems, Trans. IEICE, Vol.E73, No.9, 1990.

[28] Wilkinson, R.I., Theories for Toll Traffic Engineering in the U.S.A., BSTJ, Vol.35, No.2, 1958.

[29] Rapp, Y., Planning of Junction Network in a Multi-Exchange Areas, Ericsson Tech., Vol.20, No.1, 1964.

[30] Kuczura, A., The Interrupted Poisson Process as an Overflow Process, BSTJ, Vol.52, No.3, 1973.

[31] Akimaru, H., Takahashi, H., An Approximate Formula for Individual Call Losses in Overflow Systems, IEEE Trans. COM-31, No.6, 1983.

[32] Akimaru, H., Tokushima, H., Nishimura, T., Derivatives of Wilkinson Formula and Their Applications to the Optimum Design of Alternative Routing Systems, 9th Intern'l Teletraffic Congr., Torremorinos, 1979.

[33] Akimaru, H., Katada H., Okuda, T., An Optimum Design of Alternative Routing System with Trunk Reservation Scheme, Trans. IEICE, Vol.72-BI, No.6, 1989. (in Japanese)

[34] Le Gall, F., Bernussou, J., Blocking Probabilities for Trunk Reservation Policy, IEEE Trans. COM-35, No.3, 1987.

[35] Hoshiai, T., Takahashi, H., Akimaru, H., Modified Diffusion Approximation of Queueing Systems, Trans. IEICE, Vol.J70-B, No.3, 1987. (in Japanese)

[36] Neuts M.F., Matrix-Geometric Solution in Stochastic Models: An Algorithmic Approach, John Hopkins University Press, 1981.

[37] Neuts, M.F., Structured Stochastic Matrices of M/G/1 Type and Their Applications, Marcel Dekker, 1989.

[38] Machihara, F., Completion Time of Service Unit Interrupted by PH-Markov Renewal Customers and its Applications, 12th Intern'l Teletraffic Congr., Torino, 1988.

[39] Niu, Z.S., Akimaru, H., Studies on Mixed Delay and Non-Delay Systems in ATM Networks, 13th Intern'l. Teletraffic. Congr., Copenhagen, 1991.

[40] Niu, Z.S., Studies on Mixed Delay and Non-Delay Systems in Telecommunications Networks, Ph.D. Dissertation, Toyohashi University of Technology, Toyohashi, Japan, 1992.

[41] Fischer, W., Meier-Hellstern, K., The MMPP Cookbook, Performance Evaluation, to appear, 1992.

[42] Heffes, H., Lucantoni, D.M., A Markov Modulated Characterization of Packetized Voice and Data Traffic and Related Statistical Multiplexer Performance, IEEE JSAC, SAC-4, No.6, 1986.

[43] Okuda, T., Akimaru, H., Sakai, M., A Simplified Performance Evaluation for Packetized Voice Systems, Trans. IEICE, Vol.E73, No.6, 1990.

[44] Akimaru, H., Okuda, T., Nagai, K., A Simplified Performance Evaluation of Bursty Multi-Class Traffic Systems in the ATM, ICC'92, Chicago, 1992.

[45] Forguson, M.J., et al, Exact Results for Nonsymmetric Token Ring Systems, IEEE Trans. COM-33, No.3, 1985.

[46] Everitt, D., Simple Approximations for Token Rings, IEEE Trans. COM-34, No.7, 1986.

[47] Choudhury, G.L., Takagi, H., et al, Comments on "Exact Results for Nonsymmetric Token Ring Systems", IEEE Trans. COM-38, No.8, 1990.

[48] Bux, W., Local Area Subnetworks: A Performance Comparison, IEEE Trans. COM-29, No.10, 1981.

[49] Supot, T., Akimaru, H., A Performance Evaluation for Heterogeneous Bus CSMA/CD LANs, Trans. IEICE, Vol.E74, No.9, 1991.

[50] Pritsker, A.A.B., Introduction to Simulation and SLAM II, John Wiley & Sons, 1986.

[51] Melamed, B., Morris, R.J.T., Visual Simulation, The Performance Analysis Workstation, IEEE Computer, Vol.18, No.8, 1985.

APPENDICES

Appendix A

Basic Teletraffic Formulas

This appendix summarizes basic teletraffic formulas in relatively simple form presented in this book. A program packege for personal computer is developed[1].

A.1 Notation

The following notation is used in general[2]. For delay systems, the FIFO (first-in first-out) discipline is assumed.

a	: Offered traffic load $(= \lambda h)$ [erl]	s	: Number of servers (trunks)
n	: Number of sources (inlets)	m	: Number of waiting positions
h	: Mean holding (service) time	ξ	: h/mean early departure time
B	: Blocking probability	W	: Mean waiting time
$M(0)$	: Waiting probability	$M(t)$	: Complementary waiting time PDF
λ	: Arrival (origination) rate	ν	: Arrival rate per idle source
μ	: Service (termination) rate	ρ	: Utilization factor $(=a/s)$
${C_a}^2$	: SCV of interarrival time	${C_s}^2$	: SCV of service time
$a^*(\theta)$	: LST of interarrival time	$b^*(\theta)$	: LST of service time

[1]TDES (Traffic Design Expert System), Copyright (C) by Haruo Akimaru

[2]SCV: squared coefficient of variation LST: Laplace-Stieltjes transform

A.2 Markovian Models

[M/M/s(0)]

$$B = \frac{\frac{a^s}{s!}}{\sum_{i=0}^{s} \frac{a^i}{i!}} \equiv E_s(a) : \quad \textit{Erlang loss (B) formula} \tag{2.26}$$

[M(n)/M/s(0)]

$$B = \frac{\binom{n-1}{s} (\nu h)^s}{\sum_{i=0}^{s} \binom{n-1}{i} (\nu h)^i} : \quad \textit{Engset loss formula} \tag{2.22}$$

[M/M/s]

$$W = M(0) \frac{h}{s-a} \tag{2.38}$$

$$M(0) = \frac{\frac{a^s}{s!} \frac{s}{s-a}}{\sum_{r=0}^{s-1} \frac{a^r}{r!} + \frac{s}{s-a}} : \quad \textit{Erlang delay (C) formula} \tag{2.35}$$

$$M(t) = M(0) e^{-(1-\rho)st/h} \tag{2.41}$$

[M/M/s(m)]

$$W = M(0) \frac{h}{s} \left(\frac{1}{1-\rho} + \frac{m\rho^m}{1-\rho^m} \right)$$

$$B = P_0 \frac{a^s}{s!} \rho^m$$

$$M(0) = P_0 \frac{a^s}{s!} \frac{1-\rho^m}{1-\rho} \tag{2.52}$$

$$M(t) = M(0) e^{-st/h} \sum_{r=0}^{m-1} \frac{1-\rho^{m-r}}{1-\rho^m} \frac{(st/h)^r}{r!}$$

$$P_0 = \left(\sum_{r=0}^{s-1} \frac{a^r}{r!} + \frac{a^s}{s!} \frac{1-\rho^{m+1}}{1-\rho} \right)^{-1}$$

[M(n)/M/s]

$$W = \frac{h}{s}\Pi_0 \binom{n-1}{s} (\nu h)^s \sum_{r=0}^{n-s-1} (r+1)(n-s-1)_r \left(\frac{\nu h}{s}\right)^r$$

$$M(0) = \Pi_0 \binom{n-1}{s} (\nu h)^s \sum_{r=0}^{n-s-1} (n-s-1)_r \left(\frac{\nu h}{s}\right)^r$$

$$M(t) = \Pi_0 \binom{n-1}{s} (\nu h)^s e^{-st/h} \sum_{r=0}^{n-s-1} (n-s-1)_r \left(\frac{\nu h}{s}\right)^r \sum_{i=0}^{r} \frac{(st/h)^i}{i!} \tag{2.49}$$

$$\Pi_0 = \left[\sum_{r=0}^{s-1} \binom{n-1}{r} (\nu h)^r + \binom{n-1}{s} (\nu h)^s \sum_{r=0}^{n-s-1} (n-s-1)_r \left(\frac{\nu h}{s}\right)^r\right]^{-1}$$

$(n)_r = n(n-1)\cdots(n-r+1)$; $(n)_r = 0$, $n < r$

[M/M/s(m, ξ)]

$$W = P_0 \frac{h}{s}\frac{a^s}{s!} \sum_{r=0}^{m-1} \frac{r+1}{\pi(r+1)} \left(\frac{a}{s}\right)^r$$

$$B = P_0 \frac{a^s}{s!}\frac{1}{\pi(m)} \left(\frac{a}{s}\right)^m$$

$$M(0) = P_0 \frac{a^s}{s!} \sum_{r=0}^{m-1} \frac{1}{\pi(r+1)} \left(\frac{a}{s}\right)^r$$

$$M(t) = P_0 \frac{a^s}{s!} \sum_{r=0}^{m-1} \frac{a^r}{r!} \sum_{i=0}^{r} (-1)^i \binom{r}{i} \frac{e^{-[s+(j+1)\xi]t/h}}{1+i\xi/s} \tag{2.48}$$

$$P_0 = \left[\sum_{r=0}^{s-1} \frac{a^r}{r!} + \frac{a^s}{s!} \sum_{r=0}^{m-1} \frac{1}{\pi(r)} \left(\frac{a}{s}\right)^r\right]^{-1}$$

$$\pi(r) = \prod_{i=0}^{r} (1 + i\xi/s)$$

[M(n)/M/s(m, ξ)]

$$W = \frac{h}{s}\Pi_0 \binom{n-1}{s} (\nu h)^s \sum_{r=0}^{m-1} \frac{(r+1)(n-1-s)_r}{\pi(r+1)} \left(\frac{\nu h}{s}\right)^r$$

$$B = \Pi_0 \binom{n-1}{s} (\nu h)^s \frac{(n-1-s)_m}{\pi(m)} \left(\frac{\nu h}{s}\right)^s$$

$$M(0) = \Pi_0 \binom{n-1}{s} (\nu h)^s \sum_{r=0}^{m-1} \frac{(n-1-s)_r}{\pi(r+1)} \left(\frac{\nu h}{s}\right)^r \tag{2.47}$$

$$M(t) = \Pi_0 \binom{n-1}{s} \left(\frac{\nu h}{\xi}\right)^s \times \sum_{r=0}^{m-1} \binom{n-1-s}{r} (\nu h)^r \sum_{i=0}^{r} (-1)^r \binom{r}{i} \frac{e^{-[s+(j+1)\xi]t/h}}{1+i\xi/s}$$

$$\Pi_0 = \left[\sum_{r=0}^{s-1} \binom{n-1}{r} (\nu h)^r + \binom{n-1}{s} (\nu h)^s \sum_{r=0}^{m-1} \frac{(n-1-s)_r}{\pi(r)} \left(\frac{\nu h}{s}\right)^r\right]^{-1}$$

A.3 Poisson Input Non-Markovian Models

[M/G/1]

$$W = \frac{\rho}{1-\rho} \frac{1+C_s{}^2}{2} h : \quad \textit{Pollaczek-Khintchine formula} \tag{3.20}$$

$$W(t) = (1-a) \sum_{j=0}^{\infty} a^j R^{*j}(t) : \quad \textit{Beneš formula} \tag{3.31}$$

$R^{*j}(t)$ represents the jth convolution of the residual distribution of holding time $R(t)$.

[M/G/1(m)]

$$W = P_0 \left[mC - a^{-1} \sum_{k=1}^{m} (m-k+1) C_k\right] h \tag{3.48}$$

$$B = P_{m+1} = 1 - \frac{C}{1+aC} \tag{3.45}$$

$$M(0) = 1 - P_0 - B \tag{3.46}$$

$$C_{j+1} = \left(C_j - p_j + \sum_{k=1}^{j} p_{j-k+1} C_k\right) p_0^{-1}, \quad C_0 = 1, \quad j = 0, 1, \cdots, m-1$$

$$P_0 = \frac{1}{1+aC}, \quad C = \sum_{j=0}^{m} C_j, \quad p_j = \lim_{z \to 0} b^{*(j)}([1-z]\ \lambda)/j!$$

[M/D/s]

$$W = \left[\sum_{r=1}^{s-1} \frac{1}{1-z_r} + \frac{a^2 - s(s-1)}{2(s-a)}\right] \frac{h}{a} \tag{3.58}$$

$$M(t) = 1 - \sum_{j=0}^{k} \sum_{r=0}^{s-1} \frac{[-\lambda(jh+x)]^{(k-j+1)s-1-r}}{[(k-j+1)s-1-r]!} Q_r e^{\lambda(jh+x)},$$

$$t = kh + x\ ;\ k = 0, 1, 2, \cdots \tag{3.62}$$

$$M(0) = 1 - (s-a)\prod_{r=1}^{s-1}\frac{1}{1-z_r} = Q_{s-1} \tag{3.64}$$

$$P_j = (-1)^{s-j+1}[1 - M(0)]\, a_{s-j}, \quad j = 0, 1, \cdots, s-1$$

$$Q_r = \sum_{j=0}^{r} P_j, \quad a_r = \frac{(-1)^r y^{(s-r)}(0)}{(s-r)!}, \quad y(z) = \prod_{k=0}^{s-1}(z - z_k)$$

z_r, $r = 1, 2, \cdots, s-1$, are complex roots ($z_r \neq 1$) of the equation,

$$1 - z^s e^{a(1-z)} = 0.$$

A.4 Renewal Input Non-Markovian Models

[GI/M/s(0)]

$$B = \left[1 + \sum_{r=1}^{s}\binom{s}{r}\prod_{i=1}^{r}\frac{1}{\phi(i\mu)}\right]^{-1} \tag{3.73}$$

[GI/M/s]

$$W = M(0)\frac{h}{s(1-\omega)} \tag{3.84}$$

$$M(0) = \frac{\Pi_s}{1-\omega} \tag{3.81}$$

$$M(t) = M(0)e^{-(1-\omega)st/h} \tag{3.86}$$

$$\Pi_s = \left[\frac{1}{1-\omega} + \sum_{r=1}^{s}\binom{s}{r}\frac{s[1 - a^*(r\mu)] - r}{[s(1-\omega) - r][1 - a^*(r\mu)]}\prod_{i=1}^{r}\frac{1}{\phi(i\mu)}\right]^{-1}$$

ω, $0 < \omega < 1$, is the root of the equation,

$$\omega = a^*([1-\omega]s\mu).$$

[H_2/G/1]

$$W = \frac{\rho(1 + C_s{}^2)}{2(1-\rho)}h + \frac{[k\lambda_1 + (1-k)\lambda_2]\rho - \lambda}{\lambda_1\lambda_2(1-\rho)} + \frac{1}{\theta_0} \tag{3.108}$$

$$M(0) = 1 - \frac{\lambda_1\lambda_2}{\lambda\theta_0}(1-\rho) \tag{3.106}$$

θ_0 is the real root of the equation,

$$\theta = \frac{z(\theta) - \sqrt{z(\theta)^2 - 4\lambda_1\lambda_2[1 - b^*(\theta)]}}{2} \tag{3.105}$$

$$z(\theta) = \lambda_1 + \lambda_2 - [k\lambda_1 + (1-k)\lambda_2]b^*(\theta).$$

[$\mathbf{E}_k$/G/1]

$$W = \frac{\rho(1+C_s{}^2)}{2(1-\rho)}h - \frac{1-C_a{}^2}{2\rho(1-\rho)}h + \sum_{i=1}^{k-1}\frac{1}{\theta_i} \tag{3.113}$$

$$M(0) = 1 - \frac{(k\lambda)^k(1-\rho)}{\lambda}\prod_{i=1}^{k-1}\frac{1}{\theta_i} \tag{3.111}$$

$k = 1/C_a{}^2$ and θ_i, $i = 1, 2, \cdots, k-1$, are $k-1$ complex roots of the equation,

$$(k\lambda)^k b^*(\theta) - (k\lambda - \theta)^k = 0.$$

[GI/G/s]

$$W = \begin{cases} \dfrac{\omega_1}{1-\omega_1}h, & s = 1 \\ \hat{M}(0)\dfrac{\omega_s}{1-\omega_s}\dfrac{h}{a}, & s \geq 2 \end{cases} \quad : \textit{Modified Diffusion Approximation} \tag{6.13}$$

$$\hat{M}(0) = \frac{\sigma_s}{1-\omega_s}\left(1 + \sum_{j=1}^{s-1}\sigma_j + \frac{\sigma_s}{1-\omega_s}\right)^{-1}, \quad s \geq 2$$

$$\sigma_j = \frac{\lambda(\omega_1 - 1)}{\beta_1}\left(\frac{\alpha_j}{\alpha_1}\right)^{\eta}\exp[-2(j-1)/K_s{}^2], \quad \omega_j = \exp(-\gamma_j), \quad j = 1, 2, \cdots, s$$

$$\alpha_j = \lambda K_a{}^2 + j\mu K_s{}^2, \quad \beta_j = \lambda - j\mu, \quad \gamma_j = 2\beta_j/\alpha_j$$

$$K_a{}^2 = \left[(1-\rho)\frac{1+\omega}{1-\omega} - 1\right]\frac{1}{\rho}, \quad K_s{}^2 = 1 + \rho(C_s{}^2 - 1)$$

$$\eta = 2a\frac{K_a{}^2 + K_s{}^2}{K_s{}^4} - 1, \quad \omega = a^*([1-\omega]s\mu)$$

A.5 Non-Renewal Input Models

[MMPP/G/1]

For 2-phase MMPP input characterized by (R, Λ) defined in (6.45),

$$W_v = W_1 + W_2 = \frac{\lambda_t h^{(2)}}{2(1-\rho)} + \frac{uh}{1-\rho} : \quad \textit{Mean virtual waiting time}$$

$$W_a = \frac{\lambda_1 W_1 + \lambda_2 W_2}{\lambda_t} = W_v + \frac{uh}{\rho} : \quad \textit{Mean actual waiting time} \tag{6.59}$$

$$W_j = \pi_j \left(W_v + u \frac{\lambda_j - \lambda_t}{G} \right), \quad j = 1, 2; \quad G = \sum_{j=1}^{2} \pi_j (\lambda_j - \lambda_t)^2 \tag{6.65}$$

$$\pi_1 = \frac{r_2}{r_i + r_2}, \quad \pi_2 = 1 - \pi_1, \quad \lambda_t = \pi_1 \lambda_1 + \pi_2 \lambda_2 \tag{6.61}$$

$$u = \frac{\lambda_1 - \lambda_2}{(1-\rho)(r_1 + r_2)^2} [r_1 P_{01}(1 - \lambda_2 h) - r_2 P_{02}(1 - \lambda_1 h)] \tag{6.62}$$

$$P_{01} = \frac{[w(z) - R_1(z) - r_2](1-\rho)}{(\lambda_2 - \lambda_1)(1-z)}, \quad P_{02} = 1 - \rho - P_{01} \tag{6.63}$$

$$w(z) = \frac{R(z) + \sqrt{R(z)^2 - 4[R_1(z) R_2(z) - r_1 r_2]}}{2}$$

$$R_j(z) = \lambda_j (1 - z) + r_j, \quad j = 1, 2; \quad R(z) = R_1(z) + R_2(z)$$

z $(0 < z < 1)$ is the root of the equation,

$$z = b^*(w(z)).$$

[PH-MRP/M/$s(m)$]

For PH-MRP input with representation (α, T, T°) defined in Subsection 6.2.1,

$$B = \lambda^{-1} \boldsymbol{p}_{s+m} T^\circ \alpha \boldsymbol{e} \tag{6.30}$$

$$W = \lambda^{-1} \sum_{i=s+1}^{s+m} (i - s) \boldsymbol{p}_i \boldsymbol{e} \tag{6.31}$$

$$\boldsymbol{p}_i = (i+1)\mu \boldsymbol{p}_{i+1} C_i^{-1}, \qquad 0 \le i \le s-1$$

$$\boldsymbol{p}_i = s\mu \boldsymbol{p}_{i+1} C_i^{-1}, \qquad s \le i \le s+m-1; \ m \ge 1 \tag{6.28}$$

$$\boldsymbol{p}_{s+m} = \boldsymbol{p}_{s+m} T^\circ \alpha C_{s+m}^{-1}, \quad m \ge 0$$

$$C_i = \begin{cases} i\mu I_r - T - i\mu C_{i-1}^{-1} T^\circ \alpha, & 0 \le i \le s-1 \\ s\mu I_r - T - s\mu C_{i-1}^{-1} T^\circ \alpha, & s \le i \le s+m \end{cases} \tag{6.29}$$

Appendix B

Programs for Erlang B Formula

B.1 Integer Number of Servers

From (2.26), we have the recurrence formula,

$$E_{s+1}(a) = \frac{aE_s(a)}{s+1+aE_s(a)} \tag{B.1}$$

which is valid for an arbitrary initial value. In successive applications of (B.1), the relative error of the result does not exceed that of the initial value. Using (B.1) with the initial value $E_0(a) = 1$, the Erlang B formula for the integer number of servers is calculated.

The BASIC program is shown in Program List 1.

B.2 Real Number of Servers

We introduce the second order *incomplete gamma function*,

$$\Gamma(x+1,a) \equiv \int_a^\infty \xi^s e^{-\xi} d\xi.$$

For an integer s, this reduces to

$$\Gamma(s+1,a) = e^{-a} s! \sum_{r=0}^{s} \frac{a^r}{r!}. \tag{B.2}$$

Then, (2.26) is extended to a real number of servers s as

$$E_s(a) = \frac{a^s e^{-a}}{\Gamma(s+1,a)} \tag{B.3}$$

for which recurrence formula (B.1) also holds.

The incomplete gamma function is calculated by using the *continued fraction*,

$$\Gamma(x+1,a) = a^{x-1}e^{-a}\left(\frac{1}{a+}\,\frac{-x}{1+}\,\frac{1}{a+}\,\frac{-x+1}{1+}\,\frac{2}{a+}\,\frac{-x+2}{1+}\cdots\right). \tag{B.4}$$

The Erlang B formula for a real number of servers is calculated as follows: Let $x = s - [s]$ be the fraction of the number of servers, and k the number of terms at which the continued fraction is truncated. Then, we can calculate $E_x(a)$ in 6 decimal places with $k = [(5/4)\sqrt{x+500} + 4/a]$. Using the recurrence formula (B.1) with the initial value $E_x(a)$, $E_s(a)$ is calculated at $s = [s] + x$. The BASIC program is shown in Program List 2.

B.3 Derivatives of Erlang B Formula

Differentiating both sides of (2.26) with respect to a or s, we have the partial derivatives,

$$\frac{\partial E_s(a)}{\partial a} = \left[\frac{s}{a} - 1 + E_s(a)\right] E_s(a) \tag{B.5}$$

$$\frac{\partial E_s(a)}{\partial s} = -\Psi_{s+1}(a)E_s(a) \tag{B.6}$$

where

$$\Psi_{s+1}(a) \equiv -\frac{\partial}{\partial s}\log E_s(a) = -\frac{\partial E_s(a)}{\partial s}/E_s(a). \tag{B.7}$$

Differentiating the both sides of (B.1) with respect to s, and using abbreviations, $\Psi_{s+1} \equiv \Psi_{s+1}(a)$ and $E_s \equiv E_s(a)$, we obtain the recurrence formula,

$$\Psi_{s+1} = (1 - E_s)\left(\Psi_s + \frac{1}{s}\right), \quad s \geq 1. \tag{B.8}$$

For an integer number s, $\Psi_{s+1}(a)$ is calculated by (B.8) with the initial value,

$$\Psi_1 = -e^a \mathrm{Ei}(-a). \tag{B.9}$$

The *exponential integral function* $\mathrm{Ei}(-x)$ is defined by

$$\mathrm{Ei}(-x) = -\int_x^\infty \frac{e^{-\xi}}{\xi}d\xi$$

which is calculated by the following formulas with error $|\epsilon| < 2 \times 10^{-7}$:

$$-xe^s\mathrm{Ei}(-x) = \frac{1 + a_1/x + a_2/x^2 + a_3/x^3 + a_4/x^4}{1 + b_1/x + b_2/x^2 + b_3/x^3 + b_4/x^4} + \epsilon(x), \quad x > 1 \tag{B.10}$$

$$-\mathrm{Ei}(-x) = \sum_{r=0}^{5} c_r x^r - \log x + \epsilon(x), \quad 0 < x \leq 1. \tag{B.11}$$

The BASIC program is shown in Program List 3, which gives the coefficients in (B.10) and (B.11).

For a real number s, letting $x = s - [s]$, the derivative of $\Gamma(x+1, a)$ is evaluated by numerical differentiation of (B.4), from which we can calculate

$$\Psi_{x+1} = \frac{\partial \Gamma(x+1,a)}{\partial x} / \Gamma(x+1,a) - \log a. \tag{B.12}$$

Using this in (B.8) as the initial value, we obtain Ψ_{s+1} for a real number s.

B.4 BASIC Program Lists

[Program List 1]

```
100 PRINT "Erlang loss formula (integer number of servers)"
110 DEFINT I,S: DEFDBL A-H,T-Z
120 '
130 *MAIN
140 INPUT "s=";S          'Number of servers (positive integer)
150 INPUT "a=";A          'Offered load (erl)
160 GOSUB*ES
170 GOSUB*PRT
180 END
190 '
200 *ES                   'Recursive calculation
210  ES=1
220   FOR I=1 TO S
230    ES=A*ES/(I+A*ES)
240   NEXT
250 Y=A*(1-ES)/S          'Calculation of trunk occupancy
260 RETURN
270 '
280 *PRT                  'Output of calculated result
290  PRINT
300  PRINT USING "s=### a=###.###### Es(a)=#.###### η = #.######";S,A,ES,Y
310 RETURN
```

[Example of calculation]

s=35 a= 25.000000 Es(a)=0.011646 η=0.705967

[Program List 2]

```
100 PRINT "Erlang loss formula (continuous number of servers)"
110 DEFINT K,N: DEFDBL A-H,T-Z
120 '
130 *MAIN
140 INPUT "s=";S          'Number of servers (positive real number)
150 INPUT "a=";A          'Offered load (erl)
160 GOSUB*XK
170 GOSUB*EX
180 GOSUB*ES
190 GOSUB*PRT
200 END
210 '
220 *XK                   'Calculation of X(fraction) and K(truncated terms)
230  N=INT(S): X=S-N
240  K=INT(5/4*SQR(X+500) + 4/A)
250 RETURN
260 '
270 *EX                   'Calculation of continued fraction
280  ESK=A
290   FOR I=K TO 1 STEP -1
300    ESK=A+(-X+I-1)/(1+I/ESK)
310   NEXT: EX=ESK/A
320 RETURN
340 *ES                   'Recursive calculation
350  ES=EX
360   FOR J=1 TO N
370    ES=A*ES/(X+J+A*ES)
380   NEXT:Y=A*(1-ES)/S   'Calculation of trunk occupancy
400 RETURN
410 '
420 *PRT                  'Output of calculated result
430  PRINT
440  PRINT USING "s=### a=###.###### Es(a)=#.###### η = #.######";S,A,ES,Y
450 RETURN
```

[Example of calculation]

```
s=21.260000 a= 12.590000 Es(a)=0.007363 η=0.587832
```

[Program List 3]

```
100 PRINT "Derivatives of Erlang B formula"
110 DEFINT I,J,S: DEFDBL A-H,L-R,T-Z
120 '
130 *MAIN
140 INPUT "s=";S          'Number of servers (positive real number)
150 INPUT "a=";A          'Offered load (erl)
160 GOSUB*PSI
170 GOSUB*DER
180 GOSUB*PRT
190 END
200 '
210 *PSI                  'Calculation of Ψ(s+1,a)
220  IF A>1 THEN GOSUB*PS1A ELSE GOSUB*PS1C
230   PSI=PS1:ES=1
240    FOR I=1 TO S
250     ES=A*ES/(I+A*ES): PSI=(1-ES)*(PSI+1/I)
260    NEXT
270 RETURN
280 '
290 *PS1A                 'Calculation of Ψ(1,a) (a>1)
300    A1= 8.5733287401#: A2=18.059016973#
310    A3= 8.6347608925#: A4= .2677737343#
320    B1= 9.5733223454#: B2=25.6329561486#
330    B3=21.0996530827#: B4= 3.9584969228#
340  PSN=1+A1/A+A2/A^2+A3/A^3+A4/A^4
350  PSD=1+B1/A+B2/A^2+B3/A^3+B4/A^4
360  PS1=PSN/PSD/A
370 RETURN
380 '
390 *PS1C                 'Calculation of Ψ(1,a) (a≤1)
400    C0= -.57721566#: C1= .99999193#
410    C2= -.24991055#: C3= .05519968#
420    C4=-9.76004E-03: C5= 1.07857E-03
430  PS1=EXP(A)*(C0+C1*A+C2*A^2+C3*A^3+C4*A^4+
          C5*A^5-LOG(A))
440 RETURN
```

```
450 '
460 *DER                    'Calculation of derivatives
470  DEA=(S/A-1+ES)*ES: DES=-PSI*ES: DSA=(S/A-1+ES)/PSI
480 RETURN
490 '
500 *PRT                    'Output of calculated result
510  PRINT
520  PRINT USING "s=### a=##.###### Es(a)=#.###### Ψ = ###.######";S,A,ES,PSI
530  PRINT
540  PRINT USING "∂Es/∂a=###.###### ∂Es/∂s=###.###### ds/da=###.######";
           DEA,DES,DSA
450 RETURN
```

[Example of calculation]

```
s= 10 a=  8.5 Es(a)=0.144608 Ψ=0.340383
∂Es/∂a= 0.046431 ∂Es/∂s= -0.049222 ds/da= 0.943286
```

Appendix C

Basis of Probability Theory

This appendix summarizes the basis of probability theory as a prerequisite for understanding this book. For proofs of theorems and derivations of formulas, refer to texts on probability and stochastic theory [1].

C.1 Events and Probability

C.1.1 Events

A phenomenon occurring randomly such as call origination or termination, is called an *event*, and denoted by

$$E = \{\text{A call originates.}\}. \tag{C.1}$$

The following events are defined:

$$\begin{aligned}
&\textit{Sum Event} &&: \ A \cup B = \{A \text{ or } B\} \\
&\textit{Product Event} &&: \ A \cap B = \{A \text{ and } B\} \\
&\textit{Negative Event} &&: \ A^c = \overline{A} = \{\text{not } A\} \\
&\textit{Empty Event} &&: \ \phi = \{\text{never occurring}\} \\
&\textit{Universal Event} &&: \ \Omega = \{\text{always occurring}\} = \phi^c.
\end{aligned} \tag{C.2}$$

If events A and B do not occur simultaneously, that is

$$A \cap B = \phi, \tag{C.3}$$

[1] For example, Çinlar, E., Introduction to Stochastic Process, Prentice-Hall, 1975.

A and B are said to be *exclusive*. In this case, the sum event is expressed as

$$A \cup B = A + B \tag{C.4}$$

which is extended for exclusive n events A_i, $i = 1, 2, \cdots, n$, as

$$\bigcup_{i=1}^{n} A_i = \sum_{i=1}^{n} A_i.$$

If A occurs then B always occurs, it is said that B *includes* A (or A is included in B), and denoted by

$$A \subset B. \tag{C.5}$$

If $A \subset B$ and $B \subset A$, A is *equal to* B, and denoted by

$$A = B.$$

C.1.2 Probability

A real value $P\{A\}$ associated with an event A, is called *probability* of A, if it satisfies the following axiom:

[T0] *Kolmogorov Axiom*

(1) $P\{A\} \geq 0$.

(2) $P\{\Omega\} = 1$.

(3) If events A_i, $i = 1, 2, \cdots$, are exclusive ($A_i \cap A_j = \phi$, $i \neq j$), then

$$P\left\{\sum_{i=1}^{\infty} A_i\right\} = \sum_{i=1}^{\infty} P\{A_i\} : \quad \textit{Completely additive.}$$

[T1] *Basic Properties of Probability*

(1) For the empty event ϕ, and an event A,

$$P\{\phi\} = 0, \quad 0 \leq P\{A\} \leq 1.$$

(2) For an event A,

$$P\{A^c\} = 1 - P\{A\}.$$

(3) If $A \subset B$, then

$$P\{A\} \leq P\{B\}.$$

(4) *Additive Theorem*: For events A and B,

$$P\{A \cup B\} = P\{A\} + P\{B\} - P\{A \cap B\}.$$

In particular, if A and B are exclusive, then

$$P\{A + B\} = P\{A\} + P\{B\}.$$

(5) *Equal Probability Theorem*: If events A_i, $i = 1, 2, \cdots, n$, are exclusive, and *exhaustive* ($A_1 + A_2 + \cdots + A_n = \Omega$), and if they occur equally likely, then

$$P\{A_i\} = \frac{1}{n}, \quad i = 1, 2, \cdots, n.$$

(6) *Law of Large Number*: Letting n_A be the number of occurrences of event A in n trials, the *relative frequency*, n_A/n tends to the probability of A with a large n, that is

$$\frac{n_A}{n} \to P\{A\}, \quad n \to \infty.$$

C.1.3 Conditional Probability

For events A and B with $P\{A\} > 0$, the *conditional probability* of B, *given* A, is defined by

$$P\{B|A\} = \frac{P\{A \cap B\}}{P\{A\}}. \tag{C.6}$$

[T2] *Properties of Conditional Probability*

(1) *Product Theorem*: For events A and B,

$$P\{A \cap B\} = P\{A\}P\{B|A\} = P\{B\}P\{A|B\}.$$

(2) *Total Probability Theorem*: If events A_i, $i = 1, 2, \cdots, n$, are exclusive and exhaustive, then for an event B,

$$P\{B\} = \sum_{i=1}^{n} P\{A_i\}P\{B|A_i\}.$$

(3) *Bayes' Theorem*: With the same condition above,

$$P\{A_k|B\} = \frac{P\{A_k\}P\{B|A_k\}}{\sum_{i=1}^{n} P\{A_i\}P\{B|A_i\}}.$$

(4) *Independence*: Events A and B are said to be (*mutually*) *independent*, if one of the following conditions holds:

$$P\{A \cap B\} = P\{A\}P\{B\}, \quad P\{A|B\} = P\{A\}, \quad P\{B|A\} = P\{B\}.$$

C.2 Distribution Functions

C.2.1 Random Variables

For example, letting N be the number of calls existing in the system, the probability that j calls exist, is expressed by $P\{N = j\}$. As another example, the probability that the service time T is no greater than t, is $P\{T \leq t\}$. A real value, such as N and T, for which the probability can be defined, is called a *random variable.* The random variable is classified as *discrete* or *continuous*, according to the values it takes. Usually, a random variable is denoted by an upper case letter.

The probability that a random variable X is no greater than x, is a function of x, and is denoted as

$$F(x) = P\{X \leq x\}. \tag{C.7}$$

This is called the (*probability*) *distribution function* (PDF) of X. The *complementary distribution function* is defined by

$$F^c(x) = P\{X > x\} = 1 - F(x).$$

If X is a continuous random variable, and $F(x)$ is differentiable,

$$f(x) = \frac{dF(x)}{dx} \tag{C.8}$$

is called the (*probability*) *density function* (pdf). If X is a discrete random variable,

$$P_j = P\{X = j\} \tag{C.9}$$

is called the *probability function* (PF). The set $\{P_j\}$ is sometimes called the *probability distribution.*

[T3] *Basic Property of Distribution Function*

(1) A distribution function $F(x)$ is monotone and non-decreasing, with $F(-\infty) = 0$, $F(\infty) = 1$, and

$$0 \leq F(x) \leq 1, \quad -\infty < x < \infty.$$

(2) For $a < b$,

$$P\{a < X \leq b\} = F(b) - F(a).$$

(3) *Right continuity*: If $F(x)$ is discontinuous at $x = a$, then

$$F(a) = F(a-0) + P\{X = a\}.$$

(4) According to whether X is discrete or continuous,

$$F(x) = \sum_{j \leq x} P\{X = j\}, \quad F(x) = \int_{-\infty}^{x} f(\xi)d\xi.$$

C.2.2 Joint Distribution Function

For random variables X and Y, the *joint distribution function* $F(x,y)$ is defined by

$$F(x,y) = P\{(X \leq x) \cap (Y \leq y)\}.$$

The *marginal distribution functions* for X, and for Y, are defined by, respectively,

$$F_1(x) = F(x,\infty), \quad F_2(y) = F(\infty, y).$$

The *joint probability function* P_{jk}, and *joint density function* $f(x,y)$ are defined by,

$$\begin{aligned} P_{jk} &= P\{(X = j) \cap (Y = y)\}, \quad \text{Discrete} \\ f(x,y) &= \frac{\partial^2 F(x,y)}{\partial x \partial y}, \quad \text{Continuous.} \end{aligned} \tag{C.10}$$

The *marginal probability functions* $P_{1,j}$ and $P_{2,k}$, as well as the *marginal density functions* $f_1(x)$ and $f_2(y)$ are defined by

$$\begin{aligned} P_{1,j} &= \sum_k P_{jk}, \quad P_{2,k} = \sum_j P_{jk}, \quad \text{Discrete} \\ f_1(x) &= \int_{-\infty}^{\infty} f(x,y)dy, \quad f_2(y) = \int_{-\infty}^{\infty} f(x,y)dx, \quad \text{Continuous.} \end{aligned}$$

Random variables X and Y are said to be *independent*, if one of the following conditions holds:

$$F(x,y) = F_1(x)F_2(y), \quad P_{jk} = P_{1,j}P_{2,k}, \quad f(x,y) = f_1(x)f_2(y). \tag{C.11}$$

C.2.3 Convolution

If X and Y are independently distributed following $F_1(x)$ and $F_2(y)$, respectively, then the distribution function $G(z)$ of $Z = X + Y$ is called the *convolution* of X and Y. This is denoted by $G = F_1 * F_2$, and expressed using the *Laplace-Stieltjes integral* as

$$G(z) = \int_{-\infty}^{\infty} F_1(z-y)dF_2(y) = \int_{-\infty}^{\infty} F_2(z-x)dF_1(x). \tag{C.12}$$

Here, for a uni-valiate continuous function $\phi(\cdot)$, the Laplace-Stieltjes integral is defined by

$$\int_a^b \phi(x)dF(x) \equiv \begin{cases} \sum_{a \le j \le b} \phi(j)P_j, & \text{Discrete} \\ \int_a^b \phi(x)f(x)dx, & \text{Continuous.} \end{cases} \tag{C.13}$$

Using this, a unified expression is attained regardless of the attribute of the random variable. The distribution function and density (probability) function are expressed as

$$F(x) = \int_{-\infty}^{x} dF(\xi), \quad dF(x) = \begin{cases} P\{X = x\}, & \text{Discrete} \\ f(x)dx, & \text{Continuous.} \end{cases} \tag{C.14}$$

C.3 Expectation and Transforms

C.3.1 Expectation

Letting $\phi(X)$ be a function of random variable X, the *expectation* of $\phi(X)$ is defined by

$$E\{\phi(X)\} = \int_{-\infty}^{\infty} \phi(x)dF(x). \tag{C.15}$$

Examples are:

Mean	: $E\{X\} = \int_{-\infty}^{\infty} x dF(x) = \mu$
kth moment	: $E\{X^k\} = \int_{-\infty}^{\infty} x^k dF(x) = \mu_k$
kth central moment	: $E\{(X-\mu)^k\} = \int_{-\infty}^{\infty} (x-\mu)^k dF(x) = m_k$
Variance	: $V\{X\} = m_2 = \mu_2 - \mu^2 = \sigma^2$
Standard deviation	: $\sigma = \sqrt{V\{X\}}$
Third central moment	: $T\{X\} = m_3 = \mu_3 - 3\mu\mu_2 + 2\mu^3$
Skewness	: $S_k = T\{X\}/\sigma^3$

In particular for a random variable X non-negative integer type,

kth factorial moment : $E\{X(X-1)\cdots(X-k+1)\} = M_k$

kth binomial moment : $E\left\{\begin{pmatrix} X \\ k \end{pmatrix}\right\} = B_k = \dfrac{M_k}{k!}$.

Letting μ_X and μ_Y be the means, and σ_X and σ_Y the standard deviations of X and Y, respectively,

Covariance : $C\{X,Y\} = E\{(X-\mu_X)(Y-\mu_Y)\}$

Correlation coefficient : $\sigma_{XY} = C\{X,Y\}/(\sigma_X \sigma_Y)$.

[T4] *Properties of Expectation*

(1) For random variables X and Y,

$$E\{X+Y\} = E\{X\} + E\{Y\}$$

$$E\{XY\} = E\{X\}E\{Y\} + C\{X,Y\}$$

$$V\{X+Y\} = V\{X\} + V\{Y\} - 2C\{X,Y\}.$$

(2) In particular, if X and Y are independent,

$$C\{X,Y\} = 0$$

$$E\{XY\} = E\{X\}E\{Y\}$$

$$V\{X+Y\} = V\{X\} + V\{Y\}.$$

(3) *Non-biased estimate*: Letting $X_1, X_2, \cdots, X_n$ be samples of X, then for $n \to \infty$,

$$\text{Sample mean : } \overline{X}_n = \frac{X_1 + X_2 + \cdots + X_n}{n} \quad \to \quad E\{X\}$$

$$\text{Sample variance : } V_n = \frac{1}{n-1}\sum_{i=1}^{n}(X_i - \overline{X}_n)^2 \quad \to \quad V\{X\}.$$

(4) For a random variable X of non-negative integer type,

$$E\{X\} = M_1, \quad V\{X\} = M_2 + M_1 - {M_1}^2$$

$$T\{X\} = M_3 - (M_1 - 1)(3M_2 - 2{M_1}^2 + M_1).$$

(5) *Inversion formula*: For X of the same type as above,

$$P\{X = j\} = \sum_{k=j}^{\infty}(-1)^{k-j}\begin{pmatrix} k \\ j \end{pmatrix} B_k.$$

C.3.2 Laplace-Stieltjes Transform

For a random variable $X \geq 0$ with distribution function $F(x)$, the *Laplace-Stieltjes transform* (LST) of $F(x)$ is defined by

$$f^*(\theta) = E\{e^{-\theta X}\} = \int_0^\infty e^{-\theta x} dF(x). \tag{C.16}$$

For continuous random variables, the LST of PDF is equivalent to the *Laplace transform* (LT) of the pdf $f(x)$,

$$f^*(\theta) = \int_0^\infty e^{-\theta x} f(x) dx. \tag{C.17}$$

The term LST of X is often used interchangeably with the LST of $F(x)$.

[T5] *Formulas Related to LST*

(1) *Moment formula*:

$$E\{X^k\} = (-1)^k \lim_{\theta \to 0} \frac{d^k}{d\theta^k} f^*(\theta) = (-1)^k f^{*(k)}(0).$$

(2) *Convolution*: Letting $f_i^*(\theta)$, $i = 1, 2$, be the LST of X_i, which are independent of each other,

$$\text{LST of } Y = X_1 + X_2 : \; g^*(\theta) = f_1^*(\theta) f_2^*(\theta)$$

$$\text{LST of } Z = X_1 - X_2 : \; h^*(\theta) = f_1^*(\theta) f_2^*(-\theta).$$

(3) *Differentiation formula*: Letting $f^*(\theta)$ be the LST of $F(x)$, the LST of $f(x) = \dfrac{dF(x)}{dx}$ is given by

$$\hat{f}^*(\theta) = \theta f^*(\theta) - f(0).$$

(4) *Integration formula*: Letting $\hat{f}^*(\theta)$ be the LST of $f(x)$, the LST of $F(x) = \int_0^x f(\xi) d\xi$ is given by

$$f^*(\theta) = \hat{f}^*(\theta)/\theta.$$

(5) If $f^*(\theta)$ is the LT of $f(x)$,

$$\textit{Initial value theorem} \quad : \quad \lim_{x \to 0} f(x) = \lim_{\theta \to \infty} \theta f^*(\theta)$$

$$\textit{Final value theorem} \quad : \quad \lim_{x \to \infty} f(x) = \lim_{\theta \to 0} \theta f^*(\theta).$$

(6) *Characteristic function (CF)*: For a random variable $-\infty < X < \infty$, the CF is defined by

$$\Phi(\theta) = E\left\{e^{j\theta X}\right\} = \int_{-\infty}^{\infty} e^{j\theta x} dF(x), \quad j = \sqrt{-1} \tag{C.18}$$

which has similar properties to those of LST.

(7) *L'Hospital Theorem*:

$$\lim_{x \to a} \frac{g(x)}{h(x)} = \lim_{x \to a} \frac{g'(x)}{h'(x)} \tag{C.19}$$

which is useful for estimating the limit of indefinite form, $0/0$ or ∞/∞ at $x = a$.

C.3.3 Probability Generating Function

Letting P_j be the probability function of a non-negative integer type random variable X, the *(probability) generating function* (PGF) is defined by

$$g(z) = \sum_{j=0}^{\infty} z^j P_j. \tag{C.20}$$

[T6] *Formulas for Generating Function*

(1) *Factorial Moment formula*:

$$M_k = \lim_{z \to 1} \frac{d^k}{dz^k} g(z) = g^{(k)}(1).$$

(2) *Convolution*: Let $g_i(z)$, $i = 1, 2, \cdots$, be the PGF of X_i, which are independent of each other. Then, the PGF of $X = X_1 + X_2 + \cdots + X_n$ is given by

$$g(z) = \prod_{i=1}^{n} g_i(z).$$

(3) *Inversion formula*: Letting $g(z)$ be the PGF of X, its probability function is given by

$$P_j = \frac{1}{j!} \lim_{z \to 0} \frac{d^j g(z)}{dz^j} = \frac{1}{j!} g^{(j)}(0).$$

(4) Letting $f^*(\theta)$ be the LST of X, and $g(z)$ its PGF,

$$f^*(\theta) = g(e^{-\theta}).$$

C.4 Examples of Probability Distributions

C.4.1 Probability Distribution

A distribution function, density function or probability function characterizing a random variable, is called a *probability distribution*. The probability distribution is calssified in discrete and continuous type according to the attribute of the random variable.

Examples of probability distributions appearing in the teletraffic theory, are shown in Table C.1 for discrete type, and Table C.2 for continuous type.

C.4.2 Discrete Distributions

Table C.1 Discrete Distributions

Distributions	PF	Mean	Variance	PGF
Binomial $B(n,p)$	$\binom{n}{x} p^x q^{n-x}$ $x = 0, 1, \cdots, n$	np	npq	$(q+pz)^n$ $q = 1-p$
Geometric $Ge(p)$	pq^{x-1} $x = 1, 2, \cdots$	$\frac{1}{p}$	$\frac{q}{p^2}$	$\frac{pz}{1-qz}$
Pascal $Pas(n,p)$	$\binom{x-1}{x-n} p^x q^{x-n}$ $x = n, n+1, \cdots$	$\frac{n}{p}$	$\frac{nq}{p^2}$	$\left(\frac{pz}{1-qz}\right)^n$
Poisson $Po(\lambda)$	$\frac{\lambda^x}{x!} e^{-\lambda}$ $x = 0, 1, 2, \cdots$	λ	λ	$e^{-\lambda(1-z)}$

(1) The *binomial distribution* $B(n,p)$ tends to the *Poisson distribution* $Po(\lambda)$, when $n \to \infty$ and the mean $\lambda = np$ is kept fixed.

(2) The convolution of binomial distributions $B(n_1, p)$ and $B(n_2, p)$ becomes $B(n_1+n_2, p)$.

(3) The convolution of Poisson distributions $Po(\lambda_1)$ and $Po(\lambda_2)$ becomes $Po(\lambda_1 + \lambda_2)$.

(4) The n-fold convolution of the *geometric distribution* $Ge(p)$ becomes the *Pascal (negative binomial) distribution* $Pas(n,p)$.

(5) The convolution of Pascal distributions $Pas(n_1, p)$ and $Pas(n_2, p)$ becomes $Pas(n_1 + n_2, p)$.

C.4.3 Continuous Distributions

Table C.2 Continuous Distributions

Distributions	pdf	Mean	Variance	LST
Exponential $\mathrm{Ex}(\mu)$	$\mu e^{-\mu x}$ $x \geq 0$	$\frac{1}{\mu}$	$\frac{1}{\mu^2}$	$\frac{\mu}{\theta+\mu}$
k-Erlangian $\mathrm{E}_k(\mu)$	$\frac{(k\mu x)^{k-1}}{(k-1)!}k\mu e^{-k\mu x}$ $x \geq 0$	$\frac{1}{\mu}$	$\frac{1}{k\mu^2}$	$\left(\frac{k\mu}{\theta+k\mu}\right)^k$
n-Hyper-Exp. $\mathrm{H}_n(\mu_1,\cdots,\mu_2)$	$\sum_{i=0}^{n} k_i\mu_i e^{-\mu_i x}$ $x \geq 0; \sum_{i=1}^{n} k_i = 1$	$\sum_{i=1}^{n}\frac{k_i}{\mu_i}$	$2\sum_{i=1}^{n}\frac{k_i}{\mu_i^2} - \left(\sum_{i=1}^{n}\frac{k_i}{\mu_i}\right)^2$	$\sum_{i=1}^{n}\frac{k_i\mu_i}{\theta+\mu_i}$
Uniform $\mathrm{U}(a,b)$	$\frac{1}{b-a}$ $a < X < b$	$\frac{a+b}{2}$	$\frac{(b-a)^2}{12}$	$\frac{e^{-a\theta}-e^{-b\theta}}{\theta(b-a)}$
Unit $\mathrm{D}(m)$	1 $X = m$	m	0	$e^{-m\theta}$
Normal $\mathrm{N}(m,\sigma^2)$	$\frac{1}{\sqrt{2\pi}\sigma}e^{-\frac{(x-m)^2}{2\sigma^2}}$ $-\infty < x < \infty$	m	σ^2	$\exp\left[jm\theta - \frac{(\sigma\theta)^2}{2}\right]$ $j = \sqrt{-1}$

(1) The k-fold convolution of *exponential distribution* $\mathrm{Ex}(k\mu)$ becomes the phase *k-Erlangian distribution* $\mathrm{E}_k(\mu)$. If $k \to \infty$, $\mathrm{E}_k(\mu)$ tends to the *unit distribution* $\mathrm{D}(\mu^{-1})$.

(2) The mixed exponential distributions $\mathrm{Ex}(\mu_1)$ and $\mathrm{Ex}(\mu_2)$ become the 2nd order *hyper-exponential distribution* $\mathrm{H}_2(\mu_1,\mu_2)$. The mixed n different exponential distributions yield the order n hyper-exponential distribution.

(3) If X is distributed with the *normal distribution* $\mathrm{N}(m,\sigma^2)$, $X^* = (X-m)/\sigma$ is distributed with the *standard normal distribution* $\mathrm{N}(0,1)$.

(4) The convolution of normal distributions $\mathrm{N}(m_1,\sigma_1^2)$ and $\mathrm{N}(m_2,\sigma_2^2)$ becomes $\mathrm{N}(m_1+m_2,\sigma_1^2+\sigma_2^2)$. The convolution of $\mathrm{N}(m_i,\sigma_i{}^2)$, $i = 1,2,\cdots,n$, becomes $\mathrm{N}(m,\sigma^2)$ with $m = \sum_{i=1}^{n} m_i$, and $\sigma^2 = \sum_{i=1}^{n}\sigma_i{}^2$.

(5) *Central Limit Theorem*: If n independent random variables X_i, $i = 1,2,\cdots,n$, are distributed with an identical distribution with mean m and variance σ^2, then

$$X^* = \frac{(X_1+X_2+\cdots+X_n)/n - m}{\sqrt{n}\sigma}$$

tends to distribute with N(0,1), as $n \to \infty$.

(6) If X_i, $i = 1, 2, \cdots, n$, are independently distributed with N(0,1), then

$$Y = \sum_{i=1}^{k} X_i^2$$

is distributed with χ^2 (*chi-square*) *distribution* with degree of freedom k.

(7) If X and Y are independently distributed with N(0,1), and the χ^2 distribution with degree of freedom k, respectively, then

$$Z = X\sqrt{\frac{k}{Y}}$$

is distributed with t *distribution* with degree of freedom k.

C.5 Stochastic Processes

C.5.1 Markov Process

A set of random variables $X(t) \in E$,

$$\{X(t), t \geq 0\} \tag{C.21}$$

is called *stochastic process*. The set E is refered to as the *state space*, and the parameter t is often regarded as time.

A stochastic process is called a *Markov process*, if it has the Markov property, that is the stochastic behavior of the process in the future is only dependent on the present state, but independent of the past progress. (See Subsection 1.3.1.) This is expressed formally as

$$P\{X(t_n + s) \leq x | X(t_1) = x_1, X(t_2) = x_2, \cdots, X(t_n) = x_n\}$$
$$= P\{X(t_n + s) \leq x | X(t_n) = x_n\}, \quad s > 0; \; 0 \leq t_1 < t_2 < \cdots < t_n. \tag{C.22}$$

If the state space E is discrete, the process is called the *Markov chain*. If $X(t) = j$, the process is said to be in (visit, enter, etc.) *state* j at time t. If the process in state i can visit state j (even with multiple steps), it is said that state i is *reachable* to state j. The state space E is said to be *closed* if any state in E is not reachable to a state outside E. If E is closed and all states in E are reachable by each other, the Markov chain is said to be *irreducible*, otherwise *reducible*.

[T7] *Classification of State and Markov Chain*

(1) A state in a Markov chain is classified as:

(a) *Recurrent*, if the process leaves the state and eventually returns to that state (with probability 1).

(b) *Transient*, if the process may not return to that state (returns with probability less than 1).

(c) *Absorbing*, if the process entering that state can not visit any other state (not reachable from that state).

(2) If recurrent, the returning time is called *recurrence time*, and the state is further classified as:

(a) *Positive recurrent*, if the mean recurrence time is finite.

(b) *Null recurrent*, if the mean recurrence time is infinite.

(c) *Periodic*, if the returning time has a period. Otherwise, *Aperiodic*.

(d) *Ergodic*, if positive recurrent and aperiodic.

(3) For an irreducible Markov chain, either

(a) All states are positive recurrent, and ergodic if also aperiodic,

(b) All states are null recurrent, or

(c) All states are transient.

(d) If periodic, all states have the same period.

(4) A Markov chain is classified according to the class of all states involved. For example, if all states are ergodic as in (3-a) above, the Markov chain is called ergodic.

C.5.2 Discrete-Time Markov Chain

In a discrete-time Markov chain, in which the state changes occur only at discrete time instants, set $X_n = X(t_n)$. If $X_n = i$, the process is said to be in *state* i at time (step) n. If the *transition probability*,

$$p_{ij} = P\{X_{n+1} = j | X_n = i\} \tag{C.23}$$

can be defined independently of the time n, the Markov chain is called *time homogeneous*. This is assumed in the sequel unless otherwise stated.

Letting T_j be the *sojourn time* (steps) in state j, we have

$$P\{T_j = k\} = p_{jj}{}^k(1 - p_{jj}), \quad k = 1, 2, \cdots \tag{C.24}$$

which is the geometric distribution with mean sojourn time (steps) p_{jj}^{-1}.

Define the *transition probability matrix* by

$$P = [p_{ij}] \tag{C.25}$$

and the *state probability vector* at time n by

$$\begin{aligned} \pi^{(n)} &= (\pi_0^{(n)}, \pi_1^{(n)}, \cdots) \\ \pi_j^{(n)} &= P\{X_n = j\}. \end{aligned} \tag{C.26}$$

Then, we have

$$\pi^{(n)} = \pi^{(n-1)}P$$

which yields

$$\pi^{(n)} = \pi^{(0)}P^n. \tag{C.27}$$

[T8] *Steady State Probability for Discrete-Time Markov Chain*

(1) There exists the limit,

$$\pi = \lim_{n\to\infty} \pi^{(n)}$$

independent of the initial condition $\pi^{(0)}$, if and only if the Markov chain is ergodic.

(2) If the limit π exists, it is called the *steady state probability vector*, and is uniquely determined by

$$\begin{aligned} \pi P &= \pi \\ \pi \mathbf{e} &= 1 \end{aligned} \tag{C.28}$$

where $\mathbf{e}$ is the *unit column vector* with all components equal to 1. The component expression of (C.28) is

$$\begin{aligned} \sum_{i\in E} \pi_i p_{ij} &= \pi_j, \quad j \in E \\ \sum_{j\in E} \pi_j &= 1. \end{aligned} \tag{C.29}$$

C.5.3 Continuous-Time Markov Chain

For a continuous-time Markov chain, in which the state changes occur in continuous time, the *infinitesimal generator* (*transition rate matrix*) is defined by

$$Q(t) = [q_{ij}(t)] \tag{C.30}$$

where

$$q_{ij}(t) = \lim_{\Delta t \to 0} \frac{p_{ij}(t, t+\Delta t) - p_{ij}(t,t)}{\Delta t}, \quad \sum_{j \in E} q_{ij}(t) = 0$$

$$p_{ij}(s,t) = P\{X(t) = j | X(s) = i\}, \quad p_{ii}(t,t) = 1, \quad p_{ij}(t,t) = 0, i \neq j.$$

From the Markov property, the sojourn time T_j in state j satisfies

$$P\{T_j > s+t\} = P\{T_j > s\}P\{T_j > t\}$$

which yields the exponential distribution,

$$P\{T_j \leq x\} = 1 - \exp(q_{jj}x) \tag{C.31}$$

with mean sojourn time $-q_{jj}^{-1}$.

Define the *transition probability matrix* by

$$H(s,t) = [p_{ij}(s,t)] \tag{C.32}$$

where $H(1,1) = I$, the *identity matrix* with diagonal components equal to 1. Then, we have the *Chapman-Kolmogorov* (*C-K*) *equation*,

$$H(s,t) = H(s,u)H(u,t), \quad s \leq u \leq t. \tag{C.33}$$

Differentiating (C.33) with respect to t or s, yields

$$\textit{Forward C–K equation}: \quad \frac{dH(s,t)}{dt} = H(s,t)Q(t)$$

$$\textit{Backward C–K equation}: \quad \frac{dH(s,t)}{ds} = -Q(s)H(s,t). \tag{C.34}$$

In the time homogeneous case, setting $Q = Q(t)$ and $H(t) = H(s, s+t)$, the C-K equation reduces to

$$H(s+t) = H(s)H(t) \tag{C.35}$$

which has the solution,

$$H(t) = \exp(Qt). \tag{C.36}$$

Hence, the *state probability vector* at time t, is given by

$$\boldsymbol{\pi}(t) = \boldsymbol{\pi}(0)\exp(Qt) \tag{C.37}$$

with components $\pi_j(t) = P\{X(t) = j\},\ j = 0, 1, \cdots$.

[T9] *Steady State Probability for Continuous Time Markov Chain*

(1) The steady state probability vector,

$$\boldsymbol{\pi} = \lim_{t\to\infty} \boldsymbol{\pi}(t) \tag{C.38}$$

independent of $\boldsymbol{\pi}(0)$, exists if and only if the Markov chain is ergodic.

(2) If the vector $\boldsymbol{\pi}$ exists, it is uniquely determined by

$$\boldsymbol{\pi} Q = 0, \quad \boldsymbol{\pi}\boldsymbol{e} = 1 \tag{C.39}$$

for which the component expression is

$$\sum_{i\in E} \pi_i q_{ij} = 0, \quad j \in E; \quad \sum_{j\in E} \pi_j = 1. \tag{C.40}$$

C.5.4 Birth-Death Process

A Markov chain is called a *birth-death (B-D) process* if the state changes only by one step at a time. For the continuous time case, we define

$$\textit{Birth rate}: \ \lambda_k = q_{k,k+1}$$

$$\textit{Death rate}: \ \mu_0 = 0; \quad \mu_k = q_{k,k-1}, \quad k = 1, 2, \cdots. \tag{C.41}$$

Since $q_{kk} = -(\lambda_k + \mu_k);\ q_{jk} = 0,\ |j-k| \geq 2$, from (C.30) we have the infinitesimal generator,

$$Q = \begin{bmatrix} -\lambda_0 & \lambda_0 & & 0 \\ \mu_1 & -(\lambda_1+\mu_1) & \lambda_1 & \\ & \mu_2 & -(\lambda_2+\mu_2) & \lambda_2 \\ 0 & \ddots & \ddots & \ddots \end{bmatrix} \tag{C.42}$$

which is irreducible since it has non-zero tri-diagonal components.

From (C.40), we obtain the steady state probability vector $\boldsymbol{\pi}$ with components,

$$\begin{aligned} \pi_0 &= \left[1 + \sum_{k=1}^{\infty} \frac{\lambda_0\lambda_1\cdots\lambda_{k-1}}{\mu_1\mu_2\cdots\mu_k}\right]^{-1} \\ \pi_k &= \pi_0 \frac{\lambda_0\lambda_1\cdots\lambda_{k-1}}{\mu_1\mu_2\cdots\mu_k}, \quad k = 1, 2, \cdots, \end{aligned} \tag{C.43}$$

which exists if and only if $\pi_0 > 0$.

For example, with

$$\lambda_k = \lambda, \quad k = 0, 1, \cdots; \quad \mu_k = k\mu, \ k = 0, 1 \cdots$$

we have the Poisson distribution,

$$\pi_0 = \exp\left(-\frac{\lambda}{\mu}\right); \quad \pi_k = \exp\left(-\frac{\lambda}{\mu}\right)\frac{(\lambda/\mu)^k}{k!}, \quad k = 1, 2, \cdots,$$

which always exists since $\pi_0 > 0$.

If $\mu_k = 0, \ k \in E$, the process is called a *pure-birth (P-B) process.* For example, with $\lambda_k = \lambda, \ k = 0, 1, \cdots$, we have

$$Q = \begin{bmatrix} -\lambda & \lambda & & & \mathbf{0} \\ & -\lambda & \lambda & & \\ & & -\lambda & \lambda & \\ \mathbf{0} & & & \ddots & \ddots \end{bmatrix}. \tag{C.44}$$

If the initial condition is $X(0) = 0$, from (C.37) with $\pi_0(0) = 1; \ \pi_k(0) = 0, \ k \neq 0$, it follows that

$$\pi_0(t) = \exp(-\lambda t), \quad \pi_k(t) = \exp(-\lambda t)\frac{(\lambda t)^k}{k!}, \quad k = 1, 2, \cdots. \tag{C.45}$$

In this model, no steady state exists since $\pi_0 = \lim_{t\to\infty} \pi_0(t) = 0$.

If $\lambda_k = 0, \ k \in E$, the process is called a *pure-death (P-D) process.* For example, with $\mu_k = k\mu, \ k = 0, 1, \cdots$, we have

$$Q = \begin{bmatrix} 0 & & & \mathbf{0} \\ \mu & -\mu & & \\ & 2\mu & -2\mu & \\ \mathbf{0} & & \ddots & \ddots \end{bmatrix}. \tag{C.46}$$

With the initial condition $X(0) = n$, from (C.34) we have the binomial distribution,

$$\begin{aligned} \pi_0(t) &= [1 - \exp(-\mu t)]^n \\ \pi_k(t) &= \binom{n}{k} [\exp(-\mu t)]^k [1 - \exp(-\mu t)]^{n-k}, \quad k = 1, 2, \cdots \end{aligned} \tag{C.47}$$

which degenerates to $\pi_0 = 1; \ \pi_k = 0, \ k = 1, 2, \cdots$, as $t \to \infty$.

Appendix D

Solutions to Exercises

Chapter 1

[1] (1) The traffic load carried by the trunk group is $a = 50 \times 3/60 = 2.5\,\text{erl}$. That means the traffic load carried by a trunk is $a/5 = 0.5\,\text{erl}$.

(2) The mean number of busy trunks is equal to the carried traffic $a = 2.5\,(\text{erl})$.

[2] (1) Since the number of calls originating in 12 min follows Poisson distribution with mean $12 \times 10/60 = 2/\text{min}$, the probability that 2 or more calls originate in 12 min is $1 - (1+2)e^{-2} = 0.5940$.

(2) The interarrival time is exponentially distributed with mean $60/10 = 6\,\text{min}$. Hence, the probability that the interarrival time is no greater than 6 min is $1 - e^{-1} = 0.6231$.

[3] (1) Since the departure rate is $1/(3\,\text{min})$, the probability that the service time exceeds 6 min is $e^{-6/3} = 0.1353$.

(2) The time in question is exponentially distributed with parameter $6/3 = 2/\text{min}$. Hence, the mean time is $1/2 = 0.5\,\text{min}$.

[4] (1) The mean number of telephones being used, is equal to the traffic load, $50 \times 3/60 = 2.5\,(\text{erl})$.

(2) Since the arrival rate is $\lambda = 50/(60\,\text{min})$, from the Little formula the mean waiting time is $W = 1.2 \times 60/50 = 1.44\,\text{min}$.

Chapter 2

[1] (1) The total traffic load to the toll office is $a = 0.04 \times 1000/10 = 40\,\text{erl}$. Since from (2.26) $E_{53}(40) = 0.0082 < 0.01$, 53 trunks are required.

(2) For the over-load traffic $a = 2 \times 40\,\text{erl}$, the blocking probability becomes $E_{53}(80) = 0.3582$.

[2] (1) With $\nu = 2/\text{hr}$, $h = 3\,\text{min}$, $n = 4$ and $s = 2$, the blocking probability is $B = 0.0226$ from (2.22). The offered traffic load to subscriber lines is $a = 0.3644\,\text{erl}$ from (2.24), and thus $\eta = a(1 - B)/2 = 17.81\%$.

(2) From (2.50), the mean waiting time is $W = 2.23\,\text{sec}$. The offered load is $a = 0.3632\,\text{erl}$ from (2.51). Hence, the utilization of the subscriber lines is $\rho = a/s = 18.16\%$, since the total offered load is to be carried.

[3] (1) With $s = 20$, $\lambda = 600/(8 \times 60\,\text{min}) = 1.25/\text{min}$, $h = 12\,\text{min}$, and $a = \lambda h = 15\,\text{erl}$, from (2.36) the waiting probability is $M(0) = 0.1604$. Hence, from (2.38) the mean waiting time $W = 0.3850\,\text{min}$.

(2) With FIFO, the probability in question is $M(6\,\text{min}) = 0.0132$ from (2.41). With RSO, it is $M(6\,\text{min}) = 0.0141$ from (2.42).

[4] (1) Approximate the transmission time by exponential distribution with mean $h = 1200/4800 = 0.25\,\text{sec}$. With the traffic load $a = 6 \times 0.25 = 1.5\,\text{erl}$ and $s = 2$, if the buffer size is $m = 34$ packets, from (2.52) the discarding probability is $B = 9.08 \times 10^{-6} < 10^{-5}$ for $s = 2$.

(2) With $m = 34$ packets, it follows that $W = 321\,\text{ms}$, and $M(1\,\text{sec}) = 0.0870$ from (2.52).

Chapter 3

[1] (1) Solving (3.20) for ρ, we have

$$\rho = \frac{2W}{(1 + C_s{}^2)h + 2W}. \tag{D.3.1}$$

Since the SCV of the transmission time is $C_s{}^2 = 0.5$, substituting $W = 30\,\text{sec}$ and $h = 1\,\text{min}$, the carried load is $a = \rho = 0.4\,\text{erl}$. Hence, the number of documents transmitted per hour is $\lambda = a/h = 24$.

(2) With all the documents of the fixed form, ${C_s}^2 = 0$. Therefore, from (3.20) the mean waiting time is $W = (1/3)\,\text{min} = 20\,\text{sec}$.

[2] (1) Substituting $s = 2$ into (3.58) and (3.62), yields for $t < h$:

$$\begin{aligned} W &= \left[\frac{1}{1-z_1} + \frac{a^2-2}{2(2-a)}\right]\frac{h}{a} \\ M(t) &= 1 - \frac{(2-a)(1+az_1t/h)}{1-z_1}e^{at/h} \end{aligned} \tag{D.3.2}$$

where z_1 is a real root of $z = -e^{a(1-z)/2}$ satisfying $|z_a| < 1$ which may be obtained by successively substitutions. With $\lambda = 25/\text{sec}$, $h = 2400/32\text{k} = 75\,\text{ms}$ and $a = 1.875\,\text{erl}$, $z_1 = -0.2966$ and $W = 273\,\text{ms}$. Hence, the mean response time is $T = W + h = 348\,\text{ms}$. For $t = 100 - 75 = 25\,\text{ms}$, $M(t) = 0.8533$.

(2) In this case, $h = 2400/64\text{k} = 37.5\,\text{ms}$ and therefore $a = 0.9375\,\text{erl}$. Hence, from (3.32) $W = 28\,\text{ms}$, and the mean response time is $T = 319\,\text{ms}$. For $t = 100 - 37.5 = 62.5\,\text{ms}$, $M(t) = 0.7748$.

[3] (1) For one CPU, the system is modeled by $E_2/M/1$. Substituting $s = 1$ into (3.79) and (3.81), $\Pi_1 = \omega(1-\omega)$ and $M(0) = \omega$. With $h = 1/\mu$, from (3.84) and (3.86)

$$W = \frac{\omega}{1-\omega}h, \quad M(t) = \omega e^{-(1-\omega)t/h}. \tag{D.3.3}$$

Similar to Example 3.8 but with $\lambda = 1/(20\,\text{sec})$, $\mu = 1/(15\,\text{sec})$ and $s = 1$, we obtain $\omega = 0.6771$, $W = 31.46\,\text{sec}$ and $M(60\,\text{sec}) = 0.1861$.

(2) For $s = 1$, (3.73) becomes

$$B = \alpha(\mu) = \left(\frac{ka}{1+ka}\right)^k \tag{D.3.4}$$

where $a = \lambda/\mu$. With $a = 0.75\,\text{erl}$ and $k = 2$, $B = 0.36$.

[4] (1) The packet transmission time is $h = 64 \times 8/16\text{k} = 32\,\text{ms}$ (fixed), and its LST is $\beta(\theta) = \exp(-h\theta)$. With the given parameters, (3.101) yields $k = 0.04566$, $\lambda_1 = 1.529$, and $\lambda_2 = 60.913[/\text{ms}]$. Successive substitutions of (3.105) yields $\theta_0 = 48.67/\text{sec}$. Substituting ${C_s}^2 = 0$ and $\rho = \lambda h = 0.7026$ into (3.108), $W = 743\,\text{ms}$ follows.

(2) With symmetric H_2, from (3.102) $k = 0.02609$, $\lambda_1 = 1.182$ and $\lambda_2 = 42.738/\text{ms}$. Then, with the same argument in (1), $\theta_0 = 22.53/\text{sec}$ and $W = 568\,\text{ms}$.

Chapter 4

[1] (1) With the given parameters, we have $\lambda = 0.3$/sec, $b = 2$, $h = 1$ sec, $h^{(2)} = 1\,\text{sec}^2$ and $a = 0.6$ erl. Since $b^{(2)} = 6$ when the batch size is geometrically distributed, and $b^{(2)} = 4$ if it is fixed, from (4.19) we obtain the mean waiting time $W = 3.25$ sec in geometric case, and $W = 2.0$ sec in fixed case.

(2) Applying M/D/1 model with $\lambda = 0.6$/sec, $h = 1$ sec, from (3.20) we obtain $W = 0.75$ sec. Compared with the results in (1), we see that the batch arrival significantly increases the mean waiting time.

[2] (1) Since the customer arrival rate of each class is the same, $\lambda_i = 0.01$/ms, we have $\lambda_i h_i = 0.01 \times 10 = 0.1$ erl for each i. Therefore, from (4.24), the mean waiting times are $W_1 = 3.33$, $W_2 = 4.17$, $W_3 = 5.36$, $W_4 = 7.14$ and $W_5 = 10.0$ ms.

(2) The mean waiting time for all the jobs is $(1/5)\sum_{i=1}^{5} W_i = 6$ ms.

[3] (1) From (4.56) and (4.58), we obtain $B = 0.2929$ for non-delay Poisson input, and $W = 0.0976$ sec for delay Poisson input

(2) From (4.59), we obtain $B_2 = 0.2992$ for non-delay Poisson input, and $W_1 = 0.1116$ ms for H_2 delay input.

[4] (1) Substituting N=10, $\lambda = 0.05$/sec, $h = 1$ sec, $h^{(2)} = 2 \times 1^2 = 2\,\text{sec}^2$, $\rho = 0.05$, $\mu = 0.5$/sec, $\sigma_u^2 = 0$ into (4.73) for the exhaustive model, yields $W_E = 6$ sec.

(2) Using (4.79) for the 1-limited model, we obtain $W_L = 13$ sec.

Chapter 5

[1] (1) The Erlang loss formula for a single server is $E_1(a) = a/(1+a)$ from (2.26). Using (5.9a) and noting that $z = v/b$, (5.55) is obtained.

(2) From (5.9b), we have the rth moment of the inter-overflow time X,

$$E\{X^r\} = r[k/\lambda_1^r + (1-k)/\lambda_2^r]. \tag{D.5.1}$$

Noting that $m = E\{X\}$ and $C_a{}^2 = E\{X^2\}/m^2 - 1$, we obtain (5.56).

[2] (1) Similar to the argument in Example 5.1, we have $b = 2.1458$, $v = 4.3624$, $b^* = 22.1458$, and $v^* = 24.3624$. Using (5.11), we have $a^* = 24.6927$ and $s^* = 2.6569$. By using Program 2 in Appendix B, we obtain $E_{s^*}(a^*) = 0.8969$, $E_{s_1+s^*}(a^*) = 0.00798$ and $B = 0.008895$ for $s_1 = 33$.

(2) Substituting B obtained in (1) into (5.23), we obtain $B_0' = 0.01404$ and $B_1 = 0.01215$.

[3] (1) From Rapp formula (5.11) we obtain a^* and s^*, and then determine s_1 as shown in the table below which was calculated under the condition $E_{s_1+s^*}(a^*) = 0.01E_{s^*}(a^*)$. (Verify this condition by Program 2 in Appendix B.) The system cost f obtained from (5.31) is also shown in the table. For $k = 1.5$, f takes its minimum at $s_1 = 5$, and this agrees with the optimum value obtained from Fig.5.7.

Table D.1

s_0	a^*	s^*	$E_{s^*}(a^*)$	s_1	f $k=1.5$	f $k=1.1$
0	15.000	0.0000	1.0000	23.653	35.464	<u>26.007</u>
1	14.771	0.6449	0.9591	22.806	35.209	26.087
2	12.420	1.1149	0.9277	21.975	34.962	26.173
3	13.930	1.3756	0.9080	21.156	34.733	26.271
4	13.306	1.4154	0.9013	20.361	34.542	26.397
5	12.591	1.2610	0.9073	19.617	<u>34.425</u>	26.578
6	11.864	0.9809	0.9237	18.956	34.435	26.852

(2) In the case of $k = 1.1$, the conventional method in (5.27) leads to LTC$= 5[E_1(5) - E_2(5)] = 0.7883 > 0.83/1.1 = 0.75$ erl. This means $s_0 = 2$ is the optimum value. However, as demonstrated in the table, $s_0 = 0$ gives the minimum cost.

[4] (1) With $s_0 = s_2 = r = 1$ and $s_1 = 2$, we have the following expressions from (5.37) and (5.38):

$$B_0 = P_0(2 + a_1)/a_1, \quad B_1 = P_1 a a_2/2$$

$$P_0 = \left[\frac{2a_0{}^2 + a_1{}^2}{a_0 a_1 b(2,a)} + \frac{2a_0 + a_1}{b(3,a)}\right]^{-1}, \quad P_1 = \left[1 + a_2 + \frac{a_1 a_2}{2}\right]^{-1}. \tag{D.5.2}$$

Substituting $a_0 = a_1 = 1$ erl, $a_2 = 1.5$ erl and $a = 2$ erl into (D.5.2), we obtain $B_0 = 0.3529$ and $B_1 = 0.2308$.

(2) Iteration using Table 5.6 with $w = 1.3$ and $\epsilon = 0.0001$ yields $P_{00} = 0.1765$, $P_{01} = 0.2132$, $P_{10} = 0.1397$, $P_{11} = 0.2426$, $P_{12} = 0.1176$, which converge at $n = 6$. According to (5.54), we have the exact values $B_0 = P_{11} + P_{12} = 0.3603$, $B_1 = P_{02} + P_{12} = 0.2279$. Hence, the errors of the approximate formulas are

$$B_0: \; |0.3529 - 0.3603|/0.3603 = 2.05\,\%$$

$$B_1: \; |0.2308 - 0.2279|/0.2279 = 1.27\,\%.$$

Chapter 6

[1] (1) The overflow traffic load $a = 90 \times 1/60 = 1.5\,\mathrm{erl}$. Since the transmission time is constant, we have $C_s{}^2$=0. Hence, we have $H_2/D/2$, and from (6.14) we obtain $W = 1.135\,\mathrm{min}$.

(2) In this case, the SCV of the transmission time is $C_s{}^2 = 0.5$. Hence we have $H_2/E_2/2$, and from (6.14) we obtain $W = 1.657\,\mathrm{min}$.

[2] (1) From (6.46) we have

$$R = T + T^\circ = \begin{bmatrix} -1 & 1 \\ 1/2 & -1/2 \end{bmatrix}, \quad \Lambda = T^\circ = \begin{bmatrix} 1 & 0 \\ 0 & 1/2 \end{bmatrix}. \tag{D.6.1}$$

(2) Setting $\boldsymbol{p}_i' = \boldsymbol{p}_i/p_3(2)$, from (6.28) and (6.29) we have $\boldsymbol{p}_0' = (94, 229)/13$, $\boldsymbol{p}_1' = (147, 270)/26$, $\boldsymbol{p}_2' = (59, 82)/26$, $\boldsymbol{p}_3' = (12, 13)/13$. Normalizing these, from (6.30) and (6.31) we obtain

$$B = \lambda^{-1}\boldsymbol{p}_3 T^\circ \alpha \boldsymbol{e} = 37/836 = 0.0443$$

$$W = \lambda^{-1}\boldsymbol{p}_3 \boldsymbol{e} = 25/418 = 0.0598\,\mathrm{sec}.$$

[3] (1) We have $\lambda_n = 1.3147/\mathrm{sec}$, and from (6.77) $r_1 = 0.002189$, $r_2 = 0.002050/\mathrm{sec}$, $\lambda_1 = 1.5431$, $\lambda_2 = 1.1059/\mathrm{sec}$. With $h = 0.6667\,\mathrm{ms}$, from (6.59) we obtain $W_a = 15.3088\,\mathrm{ms}$.

(2) In this case, with $\lambda_n = 5.2695/\mathrm{sec}$ and $h = 0.6667\,\mathrm{ms}$, similarly to (1), we have $W_a = 1.0048\,\mathrm{ms}$.

[4] (1) From (6.88) $W_j = 110.85\,\mu\mathrm{sec}$, $j = 1, 3, 4, 5$, and $W_2 = 100.94\,\mu\mathrm{sec}$.

(2) $W_j = 106.04\,\mu\mathrm{sec}$, $j = 1, 3, 4, 5$, and $W_2 = 114.06\,\mu\mathrm{sec}$.

Chapter 7

[1] (1) The probability distribution function $F(x)$ is given by

$$\begin{aligned} F(x) &= \int_0^x f(\xi)d\xi = x^2, \quad 0 \le x \le 1 \\ &= 1, \qquad\qquad\qquad\quad x > 1. \end{aligned}$$

Let Y be the uniform random number, the required random number X is generated by $X = F^{-1}(Y) = \sqrt{Y}$. (See Fig.D.1.)

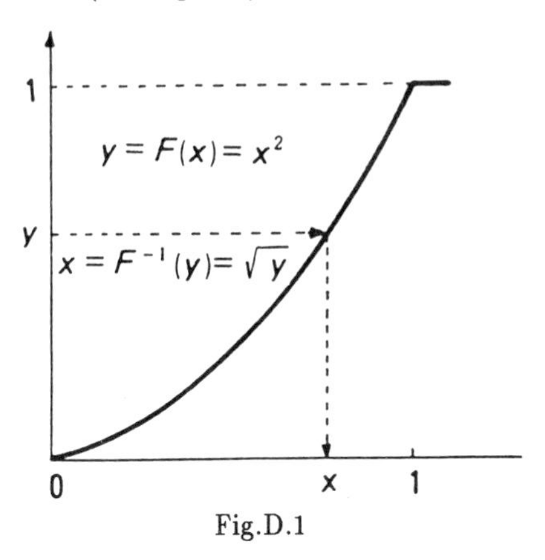

Fig.D.1

(2) The probability function $P\{X = x\}$, and distribution function $F(x)$ are given by

$$P\{X = x\} = \frac{e^{-3}3^x}{x!}, \qquad x = 0, 1, 2, \cdots$$

$$F(x) = P\{X \leq x\} = \sum_{r=0}^{x} P\{X = r\}.$$

These are shown in Table D.2. Figure D.2 illustrates $F(x)$, from which the requested random number is obtained.

Table D.2

x	$P\{X = x\}$	$F(x) = P\{X \leq x\}$
0	0.0498	0.0498
1	0.1494	0.1991
2	0.2240	0.4232
3	0.2240	0.6472
4	0.1680	0.8153
5	0.1008	0.9161
6	0.0504	0.9665
7	0.0216	0.9881
8	0.0081	0.9962
9	0.0027	0.9989
10	0.0008	0.9997

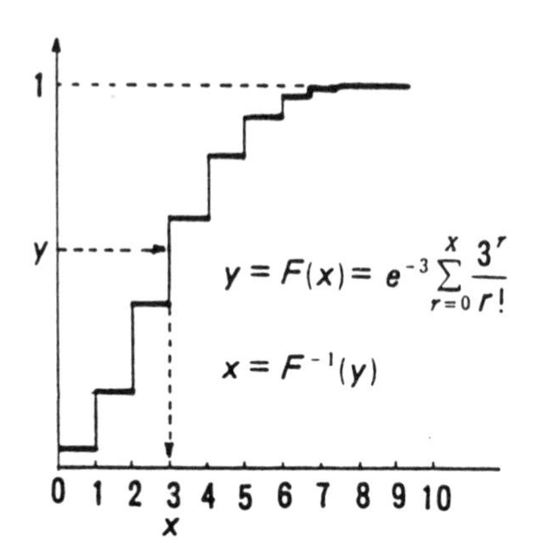

Fig.D.2

[2] (1) The mean value is $\overline{X} = 1.5944$. Setting the batch size M=10 in (7.27), we obtain the sample variance $S^2 = 0.1314$. Using the t distribution with degree of freedom $M - 1 = 9$, we have ${t_{\alpha/2}}^{M-1} = {t_{0.05/2}}^{9} = 2.262$ from Table 7.5. By substituting $u\alpha_2 = {t_{\alpha/2}}^{M-1}$ and $\sigma_m = S = 0.3625$ into (7.26), we can obtain the 95% confidence limits,

$$\left.\begin{array}{l}\mu_\alpha^+\\ \mu_\alpha^-\end{array}\right\} = 1.5944 \pm 0.2593.$$

Hence, the confidence interval is [1.3351, 1.8573].

(2) From the Little formula, the mean waiting time is given by

$$W = \overline{X}/\lambda = 1.5944/0.5 = 3.1888\,\text{sec}.$$

Index

Printing: Weihert-Druck GmbH, Darmstadt
Binding: Verlagsbuchbinderei Georg Kränkl, Heppenheim